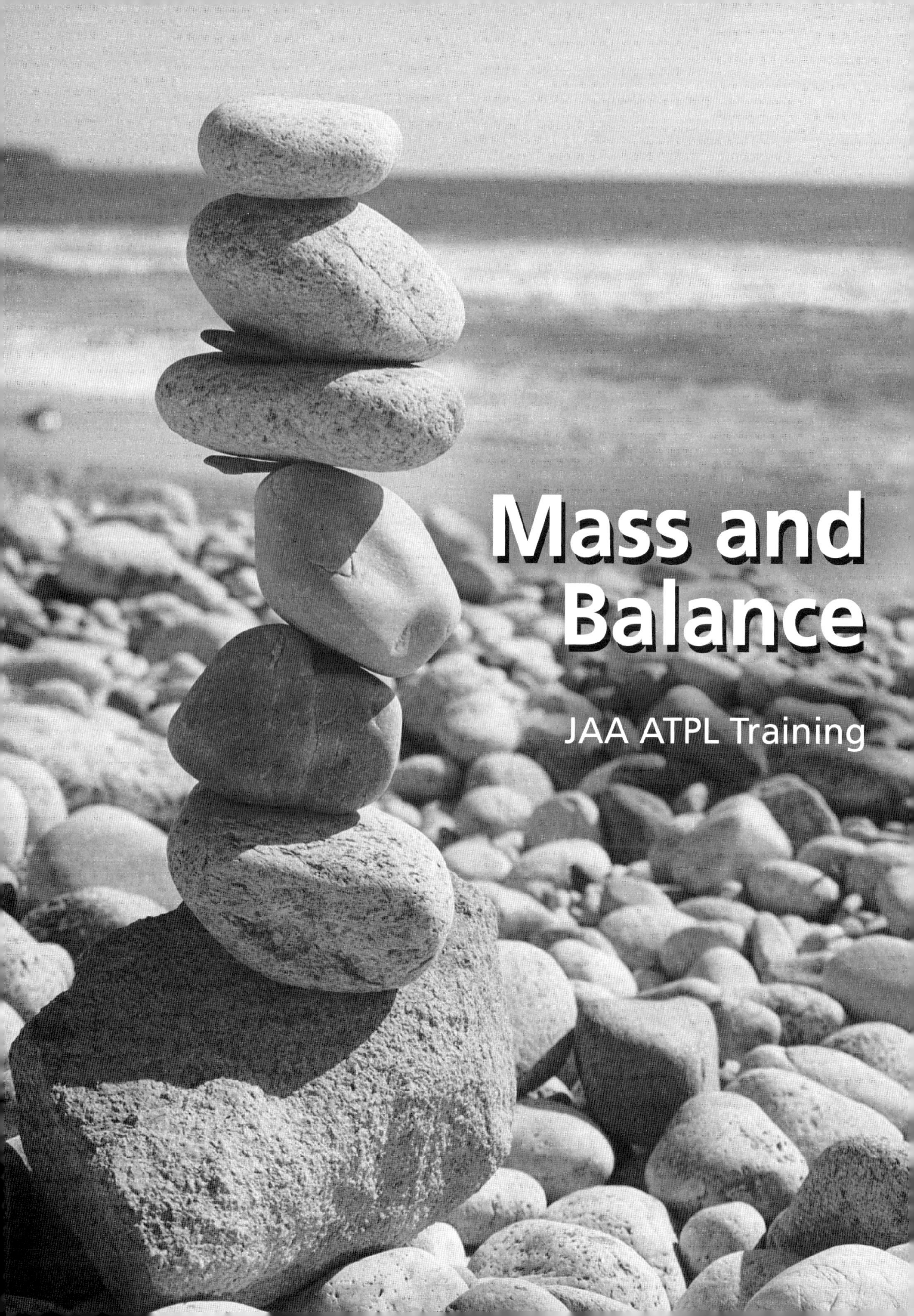

Mass and Balance

JAA ATPL Training

This book has been written and published to assist students enrolled in an approved JAA Air Transport Pilot Licence (ATPL) course in preparation for the JAA ATPL theoretical knowledge examinations. Nothing in the content of this book is to be interpreted as constituting instruction or advice relating to practical flying.

Whilst every effort has been made to ensure the accuracy of the information contained within this book, neither Jeppesen nor Atlantic Flight Training gives any warranty as to its accuracy or otherwise. Students preparing for the JAA ATPL theoretical knowledge examinations should not regard this book as a substitute for the JAA ATPL theoretical knowledge training syllabus published in the current edition of "JAR-FCL 1 Flight Crew Licensing (Aeroplanes)" (the Syllabus). The Syllabus constitutes the sole authoritative definition of the subject matter to be studied in a JAA ATPL theoretical knowledge training programme. No student should prepare for, or is entitled to enter himself/herself for, the JAA ATPL theoretical knowledge examinations without first being enrolled in a training school which has been granted approval by a JAA-authorised national aviation authority to deliver JAA ATPL training.

Contact Details:

Sales and Service Department
Jeppesen GmbH
Frankfurter Strasse 233
63263 Neu-Isenburg
Germany

Tel: ++49 (0)6102 5070
E-mail: fra-services@jeppesen.com

For further information on products and services from Jeppesen, visit our web site at:
www.jeppesen.com

ISBN 0-88487-360-9

JA310110-000

Printed in Germany

PREFACE

As the world moves toward a single standard for international pilot licensing, many nations have adopted the syllabi and regulations of the "Joint Aviation Requirements-Flight Crew Licensing" (JAR-FCL), the licensing agency of the Joint Aviation Authorities (JAA).

Though training and licensing requirements of individual national aviation authorities are similar in content and scope to the JAA curriculum, individuals who wish to train for JAA licences need access to study materials which have been specifically designed to meet the requirements of the JAA licensing system. The volumes in this series aim to cover the subject matter tested in the JAA ATPL ground examinations as set forth in the ATPL training syllabus, contained in the JAA publication, "JAR-FCL 1 (Aeroplanes)".

The JAA regulations specify that all those who wish to obtain a JAA ATPL must study with a flying training organisation (FTO) which has been granted approval by a JAA-authorised national aviation authority to deliver JAA ATPL training. While the formal responsibility to prepare you for both the skill tests and the ground examinations lies with the FTO, these Jeppesen manuals will provide a comprehensive and necessary background for your formal training.

Jeppesen is acknowledged as the world's leading supplier of flight information services, and provides a full range of print and electronic flight information services, including navigation data, computerised flight planning, aviation software products, aviation weather services, maintenance information, and pilot training systems and supplies. Jeppesen counts among its customer base all US airlines and the majority of international airlines worldwide. It also serves the large general and business aviation markets. These manuals enable you to draw on Jeppesen's vast experience as an acknowledged expert in the development and publication of pilot training materials.

We at Jeppesen wish you success in your flying and training, and we are confident that your study of these manuals will be of great value in preparing for the JAA ATPL ground examinations.

The next three pages contain a list and content description of all the volumes in the ATPL series.

ATPL Series

Meteorology (JAR Ref 050)

- The Atmosphere
- Wind
- Thermodynamics
- Clouds and Fog
- Precipitation
- Air Masses and Fronts
- Pressure System
- Climatology
- Flight Hazards
- Meteorological Information

General Navigation (JAR Ref 061)

- Basics of Navigation
- Magnetism
- Compasses
- Charts
- Dead Reckoning Navigation
- In-Flight Navigation
- Inertial Navigation Systems

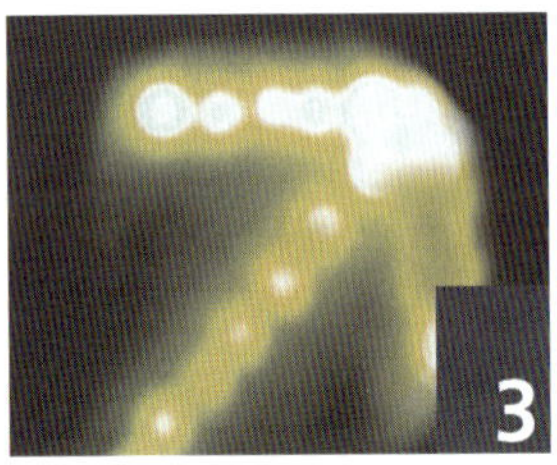

Radio Navigation (JAR Ref 062)

- Radio Aids
- Self-contained and External-Referenced Navigation Systems
- Basic Radar Principles
- Area Navigation Systems
- Basic Radio Propagation Theory

Airframes and Systems (JAR Ref 021 01)

- Fuselage
- Windows
- Wings
- Stabilising Surfaces
- Landing Gear
- Flight Controls
- Hydraulics
- Pneumatic Systems
- Air Conditioning System
- Pressurisation
- De-Ice / Anti-Ice Systems
- Fuel Systems

Powerplant (JAR Ref 021 03)

- Piston Engine
- Turbine Engine
- Engine Construction
- Engine Systems
- Auxiliary Power Unit (APU)

Electrics (JAR Ref 021 02)

- Direct Current
- Alternating Current
- Batteries
- Magnetism
- Generator / Alternator
- Semiconductors
- Circuits

Instrumentation (JAR Ref 022)

- Flight Instruments
- Automatic Flight Control Systems
- Warning and Recording Equipment
- Powerplant and System Monitoring Instruments

Principles of Flight (JAR Ref 080)

- Laws and Definitions
- Aerofoil Airflow
- Aeroplane Airflow
- Lift Coefficient
- Total Drag
- Ground Effect
- Stall
- C_{LMAX} Augmentation
- Lift Coefficient and Speed
- Boundary Layer
- High Speed Flight
- Stability
- Flying Controls
- Adverse Weather Conditions
- Propellers
- Operating Limitations
- Flight Mechanics

Performance (JAR Ref 032)

- Single-Engine Aeroplanes – Not certified under JAR/FAR 25 (Performance Class B)
- Multi-Engine Aeroplanes – Not certified under JAR/FAR 25 (Performance Class B)
- Aeroplanes certified under JAR/FAR 25 (Performance Class A)

Mass and Balance (JAR Ref 031)

- Definition and Terminology
- Limits
- Loading
- Centre of Gravity

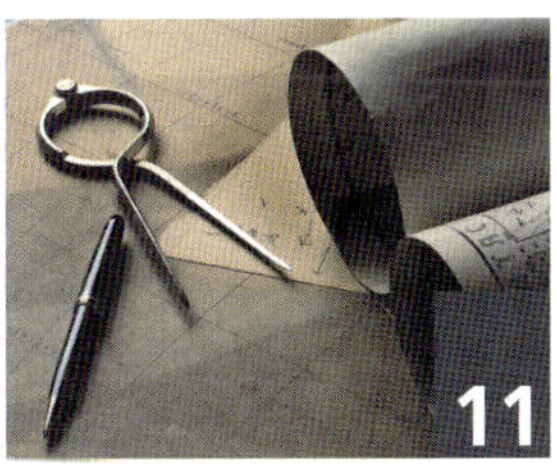

Flight Planning (JAR Ref 033)

- Flight Plan for Cross-Country Flights
- ICAO ATC Flight Planning
- IFR (Airways) Flight Planning
- Jeppesen Airway Manual
- Meteorological Messages
- Point of Equal Time
- Point of Safe Return
- Medium Range Jet Transport Planning

Air Law (JAR Ref 010)

- International Agreements and Organisations
- Annex 8 – Airworthiness of Aircraft
- Annex 7 – Aircraft Nationality and Registration Marks
- Annex 1 – Licensing
- Rules of the Air
- Procedures for Air Navigation
- Air Traffic Services
- Aerodromes
- Facilitation
- Search and Rescue
- Security
- Aircraft Accident Investigation
- JAR-FCL
- National Law

Human Performance and Limitations (JAR Ref 040)

- Human Factors
- Aviation Physiology and Health Maintenance
- Aviation Psychology

Operational Procedures (JAR Ref 070)

- Operator
- Air Operations Certificate
- Flight Operations
- Aerodrome Operating Minima
- Low Visibility Operations
- Special Operational Procedures and Hazards
- Transoceanic and Polar Flight

Communications (JAR Ref 090)

- Definitions
- General Operation Procedures
- Relevant Weather Information
- Communication Failure
- VHF Propagation
- Allocation of Frequencies
- Distress and Urgency Procedures
- Aerodrome Control
- Approach Control
- Area Control

CHAPTER 1

Introduction to Mass and Balance

CHAPTER 2

Mass and Balance Theory

CHAPTER 3

Factors Affecting Mass and Balance in Aircraft

CHAPTER 4

Mass Definitions and Limitations

CHAPTER 5

Aircraft Weighing and Floor Loading

CHAPTER 6

Load Shifting, Load Addition, and Load Subtraction

CHAPTER 7

Mean Aerodynamic Chord

CHAPTER 8

JAR-Ops 1 Requirements

CHAPTER 8 (Continued)

CHAPTER 9

Loading Manifests, SEP 1, and MEP 1

CHAPTER 10

Medium Range Jet Transport (MRJT)

A PA31 aircraft crashed just after take-off from a Caribbean island. The pilot was killed. The aircraft had been loaded with a cargo of shellfish (prawns).

Subsequent investigation revealed that the pilot had supervised the loading of the cargo into the aircraft. It comprised a number of boxes and the weight of each box was stamped on the side. He signed the load sheet. No cargo nets were used to prevent the boxes from slipping.

On the cargo load sheet with the description of the cargo and the weight set out, it did not look as if the aircraft was overloaded. The distribution of the cargo might well have been a problem since with the absence of cargo nets, on take-off the cargo could have slipped rearwards causing an aft C of G which would have led to the nose pitching up sharply and stalling the aircraft.

A further line of investigation led the team to examine the weight of cargo carried compared with the weight printed on each box. As shellfish must be 100% fresh before preparation they are transported live, to slow their metabolism they are packed in crushed ice. The weight printed on the box reflected the weight of shellfish not the total weight of the box.

The pilot had not taken account of the weight of the ice. When the investigators visited the site, of course all of the ice had melted and the water evaporated, it was some time before this explanation was discovered.

It, therefore, became apparent that the aircraft was considerably overweight when the take-off was attempted. The aircraft stalled on take-off through a combination of being overweight and the rearwards C of G. As a result of this the insurance company declined to pay the claim of the aircraft operator.

Chapter 1
Introduction to Mass and Balance

INTRODUCTION

The subject of Mass and Balance for the JAR exams deals with the loading of aircraft. This is to ensure that they are not overloaded or incorrectly loaded.

In the JAR syllabus and examinations, the subject of Mass and Balance is an integral part of Aircraft Flight Performance and Planning. The subject encompasses elements of Principles of Flight, Performance, and Flight Planning as well as the main subject of Mass and Balance.

These notes will teach the fundamentals of mass and balance, the current definitions applicable to the course, and, of course, provide preparation for the examination. They are intended to be used in conjunction with the CAP 696-JAR FCL Examinations Loading Manual. They provide direction and guidance through the manual, to facilitate familiarity with its layout and content. This makes it easier to find the relevant data and calculate answers quickly and accurately during the examination.

CAP 696-JAR FCL EXAMINATIONS LOADING MANUAL

This manual is split into 4 sections:

Section 1	General Notes	
Section 2	Data for single-engine piston/propeller aeroplane	(SEP 1)
Section 3	Data for light twin-engine piston/propeller aeroplane	(MEP 1)
Section 4	Data for medium range twin jet	(MRJT 1)

Please note that the data given in the aircraft data sheets is for examination purposes only. It is not to be used for any flight planning that involves a real aeroplane of the types shown.

SECTION 1-GENERAL NOTES (PAGES 1 TO 4)

AIRCRAFT DESCRIPTION

The aircraft descriptions are for generic types related to the classes of aircraft used in the JAR examinations. The data for each aircraft is given on different coloured paper. This colour coding is used in the sister publications for the Performance and Flight Planning examinations.

Green Paper — Single-Engine Piston

This is based on the Beech Bonanza, a single, piston-engine aeroplane that was manufactured prior to the implementation of JARs and is not certified under JAR 23 (Light Aeroplanes). As this aeroplane's MTOM is less than 5700 kg and it is piston powered, it is classed as a JAR performance class B aircraft. For the performance group of exams (mass and balance, flight planning, and performance), this is referred to as SEP 1.

Blue Paper — Multi-Engine Piston

This is based on the Piper Seneca, a multi, piston-engine aeroplane that was manufactured prior to the implementation of JARs and, therefore, is not certified under JAR 23 (Light Aeroplanes). As the MTOM is less than 5700 kg, the aeroplane is classed as performance B and is referred to as MEP 1.

White Paper — Medium-Range Jet Transport

A medium-range twin turbine-engine aeroplane certified under JAR 25 performance class A is referred to as MRJT 1.

DEFINITIONS

The main definitions are given on pages 2, 3, and 4. These are important to know during the early part of the course. This will assist in answering the questions quickly. Note that definitions in CAP 696 are given in two formats:

- If the definition is in normal text, it can be found in either ICAO or JAA documentation.
- If the definition is in italics, it is not an ICAO or JAA definition but is one that is in common use.

Throughout the text, reference is made as to which page of the CAP 696 and under what heading the item is found.

CONVERSIONS (PAGE 4)

The conversion factors are given to the 8th decimal place because they have been taken from the ICAO manual. Use only four decimal places for any calculation where a conversion is required.

For the JAA exams, all calculations should be possible using the CRP 5. However, it is highly recommended that a calculator be used in Mass and Balance. The answers indicate the level of accuracy required, whether it is a whole number or to two places decimal. In any calculation, convert using four places decimal, and then work to three places decimal, rounding up or down as required.

The following conversion factors are taken from the ICAO Annexes.

Conversions CAP Page 4		
Mass Conversions	**ICAO Conversion Factor**	**Use the following conversion**
Pounds (lb) to Kilograms (kg)	lb x 0.45359237 kg	lb x 0.4536 kg
Kilograms to Pounds	kg x 2.20462262 lb	kg x 2.2046 lb
Volumes (Liquids)		
Imperial Gallons to Litres (L)	Imp Gal x 4.546092	Imp Gal x 4.5461
US Gallons to Litres	US Gal x 3.785412	US Gal x 3.7854
Lengths		
Feet (ft) to Metres (m)	ft x 0.3048	ft x 0.3048
Distances		
Nautical Mile (nm) to metres (m)	nm x 1852.0	nm x 1852.0

Note: The last two conversions will have to be used as listed in the CAP.

Not given: 100 cm = 1 metre
1 foot = 12 inches which means 0.5 of a foot = 6 inches
1 metric tonne = 1000 kg

Chapter 1 Practice 1

1. Convert 4300 kg into pounds
2. Convert 35 ft into metres
3. Convert 5.76 m into feet

Practice 1 answers given on page 1-6.

VOLUMETRIC AND MASS CONVERSIONS

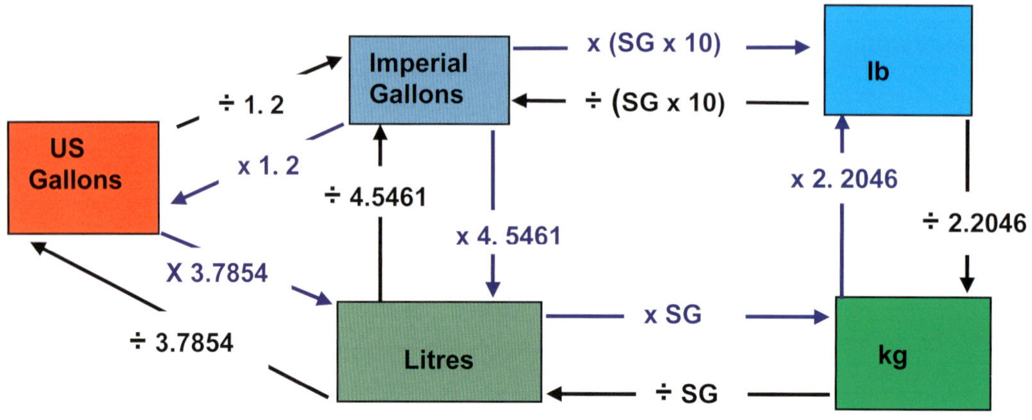

There are three volumetric measures of liquids used: the Imperial Gallon (Imp Gal), the US Gallon (US Gal), and the litre (ltr). A supplementary conversion that may be required in addition to those on Page 4 is that from Imperial Gallons to US Gallons.

The Imperial Gallon is larger than the US Gallon; there are 1.2 US Gallons to one Imperial Gallon. To convert from Imperial Gallons to US Gallons, multiply by 1.2. To convert from US Gallons to Imperial Gallons, divide by 1.2.

For example:

5 Imperial Gallons equal 6 US Gallons	5 x 1.2 = 6
5 US Gallons equal 4.17 Imperial Gallons	5 ÷ 1.2 = 4.17

Another conversion is that from liquid volume into mass. The calculation is made using the fluid's Specific Gravity or SG. This is based on the fact that 1 Imperial Gallon of water has a mass of 10 pounds. The SG is given a value of 1, therefore, the SG has a constant of 10.

For example:

An Imperial Gallon of fluid with an SG 0.72 has a mass of 7.2 pounds (0.72 x 10). As the US Gallon is only 0.8333 of an Imperial Gallon, the fastest way to find its mass is to divide the mass of an equivalent UK gallon by 1.2.

1 US Gallon of fuel at SG 0.72 equals = 6 lb (7.2 lb ÷ 1.2)
1 US Gallon at SG 0.72 has a mass of 6 pounds

A litre of water has a mass of 1 kilogram, so the Specific Gravity of 1 is used. This means that a litre of fuel at SG 0.72 has a mass of 0.72 kg or 720 grams. It is normal to use the decimal place of the kilogram.

The mass in kg per US Gallon is found by dividing the mass by the volume. The standard conversion for US Gal to kg is 1 US gal = 3.030 kg. This is found on page 22 of the CAP 696 in the bottom line of Fig 4.5.

During powered flight, fuel is consumed. This is termed **trip fuel**. The rate of this consumption depends on a combination of factors such as aircraft mass, power setting, flight level, and meteorological conditions. In light aircraft, the fuel consumption is often quoted in volume/time (e.g. US Gallons per hour). In larger aircraft, the fuel consumption is normally quoted in mass/time (e.g. kg per hour).

For example:

A twin jet engine aircraft's fuel consumption is 3000 US Gallons per hour per engine for a 4-hour flight. What is the mass of fuel that is burnt?

Using the standard value for jet fuel of 1 US Gallon = 3.030 kg.

3000 US Gallons per hour per engine	= 12 000 US Gal per engine
12 000 US Gal per engine x 2 engines	= 24 000 US Gal
24 000 US Gal x 3.030 kg	= 72 720 kg

Chapter 1 Practice 2

1. Convert 300 litres into Imperial Gallons.
2. Total 297 Imp Gal and 789 ltr and give answer in US Gal.
3. Calculate the mass in kg of 60 US Gallons of jet fuel at 3.030 kg per gallon.

Practice 2 answers given on page 1-6.

ANSWERS TO PRACTICES

PRACTICE 1

	Item	Factor	Answers
1.	Convert 4300 kg into pounds	x 2.2046 lb	**9479.78 lb**
2.	Convert 35 ft into metres	x 0.3048	**10.668 m**
3.	Convert 5.76 m into feet	÷ 0.3048	**18.898 ft**

PRACTICE 2

	Item	Factor	Answers
1.	Convert 300 litres into Imp Gal	÷ 4.5461	**65.99 Imp Gal**
2.	Total 297 Imp Gal and 789 ltr Answer in US Gal	x 1.2 ÷ 3.7854	**356.40 US Gal** **208.43 US Gal** **564.83 US Gal**
3.	Calculate the mass in kg of 60 US Gallons of jet fuel at 3.030 kg per gallon	x 3.030	**181.8 kg**

Chapter 2
Mass and Balance Theory

DEFINITIONS

MASS

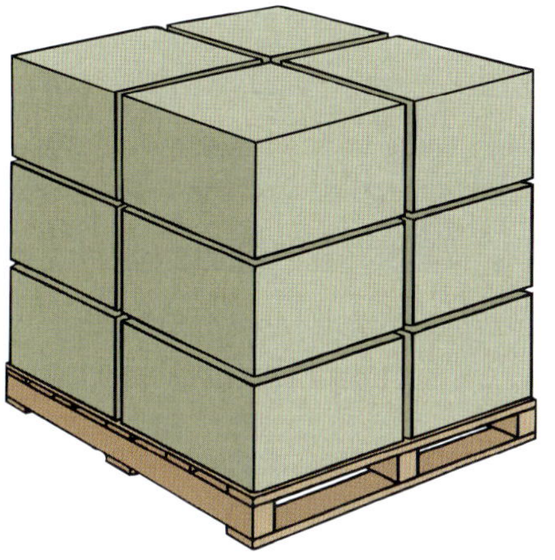

Fig. 2.1

Mass is the amount of an item inside a body (i.e. each box on this pallet has a mass, and collectively the loaded pallet has a mass). This is expressed in kilograms or pounds depending on the system used. Conventionally, it is considered that the item in question is on the Earth's surface and subjected to the force of 1g, our standard surface condition. However, when a person steps onto a set of scales, the action is termed **weight** versus **mass**.

To be scientifically correct, weight is the resultant force created by the effect of the acceleration of gravity acting on the mass of a body. This is covered later.

CENTRE OF GRAVITY (CG) — the point through which the force of gravity is said to act on a mass.

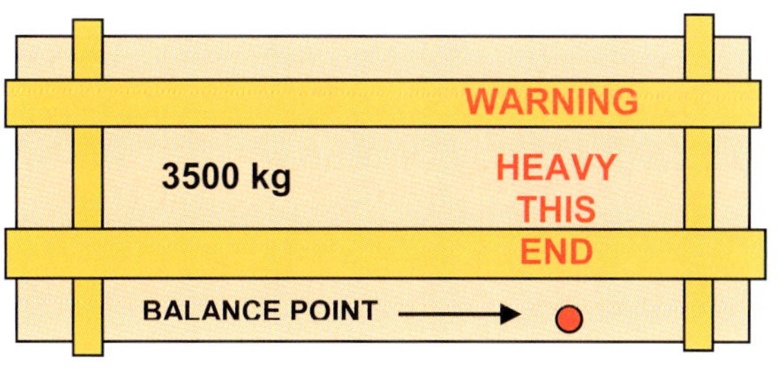

Fig. 2.2

The weight of an object acts directly toward the centre of the Earth through the item's Centre of Gravity (CG), as can be seen in fig. 2.2. The CG, which is also the item's balance point, does not necessarily occur at the mid point of the item.

BALANCE ARM (BA) — the distance from the Datum to the Centre of Gravity of a mass

MOMENT — the product of the mass and the balance arm = Mass X Arm

When holding an item at arm's length, it seems to feel heavier than if held close to the body. The mass is trying to pull the arm down. This is the turning moment and is created by multiplying the mass by the arm. The correct term for the arm is balance arm (BA). However, it is frequently referred to as the lever arm or moment arm.

Mass x Arm = Moment

The moment is a force that is acting toward the centre of the Earth and is quantified by both units of mass and units of distance. It is conventional to express this force with the unit of mass followed by the unit of distance.

For example: 300 kg in, which in some cases is written as 300 kg/in
It expresses that the force is equal to 300 kg acting over one inch or 1 kg acting over 300 in.

300 kg x 1 in = 300 kg/in
1 kg x 300 in = 300 kg/in

While this is simply a basic multiplication task, be aware that the arm's length and the mass can be given in imperial or metric units of measurement. In cases where mixed units are given, they must be converted to the same units; otherwise, large inaccuracies may occur.

Examples are given below.

Mass (weight)		**Arm**	**Moment**	**Mass (weight)**		**Arm**	**Moment**
10 kg	X	10 cm	100 kg cm	10 kg	X	10 in	100 kg in
10 lb	X	10 cm	100 lb cm	10 lb	X	10 in	100 lb in
10 kg	X	10 ft	100 kg ft	10 kg	X	10 m	100 kg m
10 lb	X	10 ft	100 lb ft	10 lb	X	10 m	100 lb m

It is also very important to correctly label the units. This can be seen from the table above, where the weights and arms are numerically the same in every case as are the resulting moments. However, the labels denote the size and, hence, effect.

Example. Calculate the total mass and moment:

Mass		Arm	Moment
10 kg	X	10 cm	100 kg cm
10 lb	X	10 cm	100 lb cm

1. Convert 10 lb into kg. Refer to the CAP, page 4 but use to 4 pd.

Mass to be Converted		Conversion	Resultant Mass
10 lb	X	0.4536 kg	= 4.536 kg

2. Now recalculate the moment using this mass.

Mass		Arm	Moment
10 kg	X	10 cm	100 kg cm
4.536 kg	X	10 cm	45.36 kg cm

PRACTICE 1

Recalculate the following to express the all moments in metric using the conversion factors as given in the CAP to 4 pd and give your answer to 3 pd.

The worked answers are given at the end of this chapter on page 2-16.

Mass (weight)		Arm	➔	Mass kg		Arm m	Moments kg/m
10 lb	X	10 cm	➔		X		
10 kg	X	10 ft	➔		X		
10 lb	X	10 ft	➔		X		

DATUM OR REFERENCE DATUM — (Relative to an aeroplane) is that (vertical) plane from which the centres of gravity of all masses are referenced.

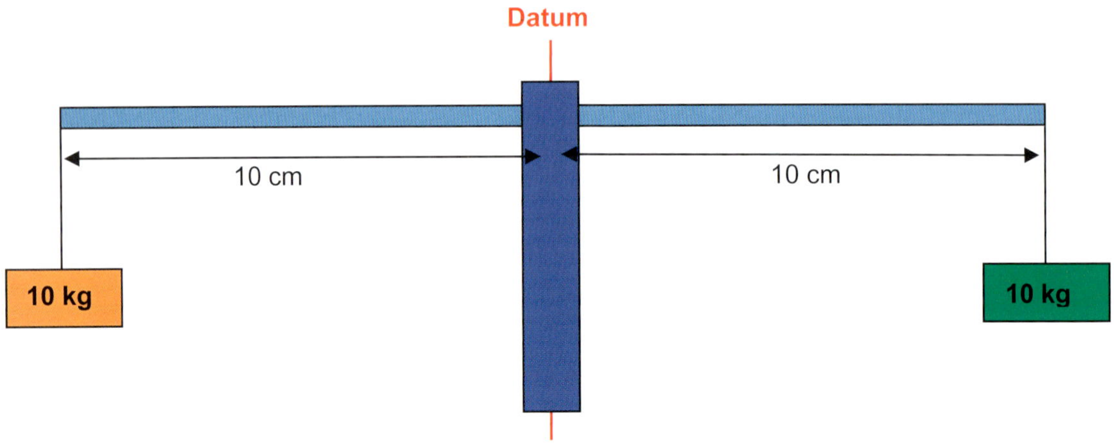

Fig. 2.3

As the moment, or to be completely correct the **turning moment**, is the product of multiplying the mass by an arm, there is a need to quantify the length of the arm by giving the location of the mass in relation to an arbitrary point. This arbitrary point is termed the reference datum, which is often shortened to just datum. It is in effect a vertical plane from which all balance arms are measured; its location is unimportant provided all arm measurements are taken from it and that the units used are the same. In fig. 2.3 of a balance scale, the datum has been placed so that it passes through the scale's pivot point. Therefore, the arms are of equal length.

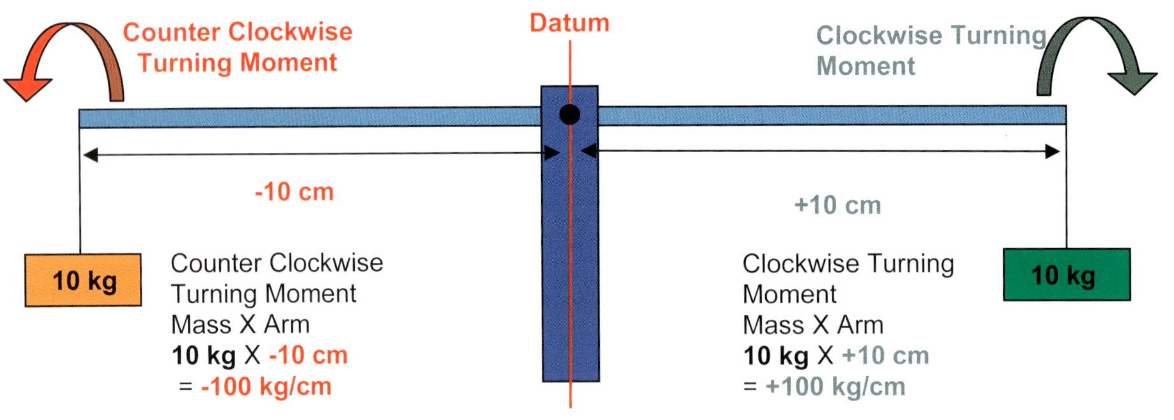

Fig. 2.4

TURNING MOMENTS

Each mass, acting over the length of its arm, produces a turning moment about the datum. In fig. 2.4, these are shown as clockwise and counter clockwise. It is conventional to label the arm to the left of the datum as negative and the arm to the right of the datum positive.

To calculate the location of the Centre of Gravity of a series of masses, use the formula:

TMo ÷ TM = CG

TMo is the total moment of all the masses involved acting over their arms from the datum.
TM is the total weight of all the masses involved.

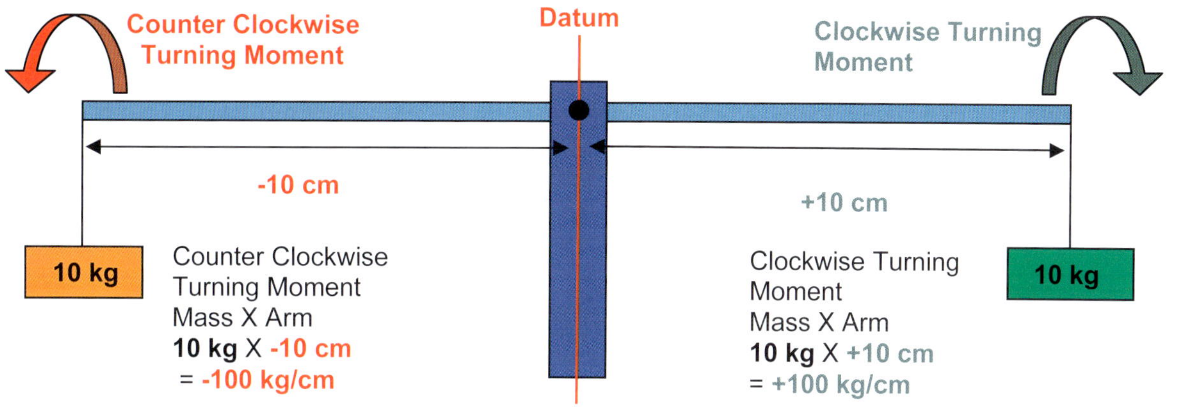

Fig. 2.5

As the arms are of equal length and the masses suspended from each arm are also equal, the two sets of moments cancel each other out. The scale is in balance with the total weight acting down through the pivot point, which, in this case, happens to be the selected datum. Whilst it is obvious that the scale in the diagram above is in balance, the calculation below that proves this is given to show the basic methodology. As the left arm moment cancels the right arm moment, it is conventional to label the left arm minus and the right arm plus:

	Mass		**Arm**	**- Moment**	**+ Moment**
Left arm	10 kg	X	-10 cm	-100 kg cm	
Right arm	10 kg	X	+10 cm		+100 kg cm
Totals	20 kg			-100 kg cm	+100 kg cm
Total Mass	**20 kg**		**Total Moment**	**0.0 kg cm**	

TMo ÷ TM = CG
Total Moment ÷ Total Mass
0.0 kg cm ÷ 20 kg = 0 cm

The datum is always labelled 0 or 0.0, so the CG is located on the datum.

If the same balance scale is used, but the location of the datum is moved as per fig. 2.6, the centre of gravity is still calculated using the **TMo ÷ TM** formula.

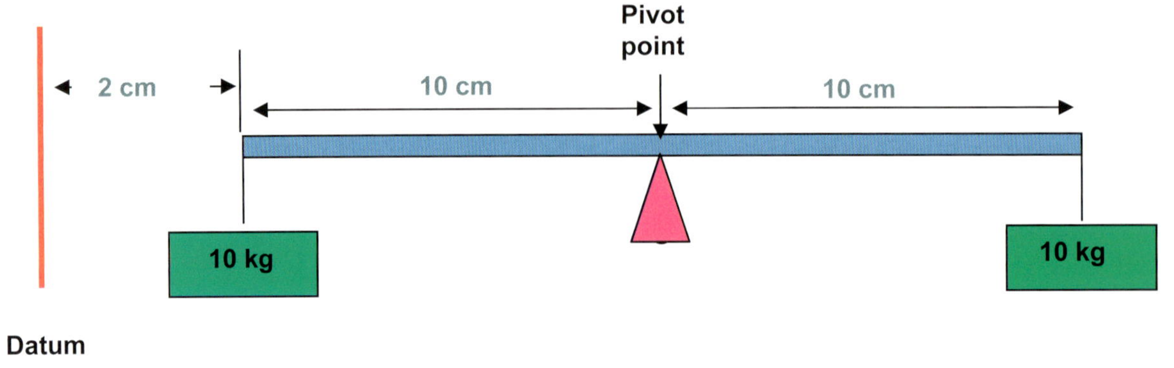

Fig. 2.6

In fig. 2.6, the datum is located 2 cm to the left of the left arm and mass. Logic still tells us that the scales are in balance as the masses and distances from the pivot point are equal. The calculation alters as both balance arms are to the right of the datum, making them both positive. The CG stays at the same point, in this case 12 cm to the right of the datum.

	Mass		B.A.	- Moment	+ Moments
	10 kg	X	+2 cm		+20 kg cm
	10 kg	X	+22 cm		+220 kg cm
Totals	**20 kg**		**Total Moment**		+240 kg cm

Centre of Gravity = TMo ÷ TM

CG = +240 kg cm ÷ **20 kg**

CG = **+12 cm**

The centre of gravity is located 12 cm to the right of the datum.

If the mass is changed then the balance point (the CG) changes.

Example

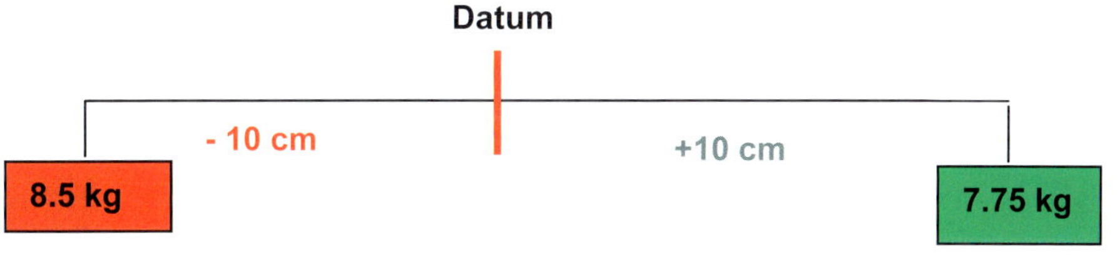

Fig. 2.7

The left balance arm has a weight of 8.5 kg 10 cm from the datum. The right arm has a mass of 7.75 kg 10 cm from the datum. The datum is located on the pivot point. Working to 3 pd and giving the answer to 2 pd, where is the CG located?

Item	**Mass**		**Arm**	**- Moment**	**+ Moments**
Left arm	8.5 kg	X	**-10 cm**	**-85 kg cm**	
Right arm	7.75 kg	X	**+10 cm**		**+77.5 kg cm**
Totals	**16.25 kg**			**-85 kg cm**	**+77.5 kg cm**
			Total Moment	**-7.5 kg cm**	

Calculate the CG position

CG = **-7.5 kg cm** ÷ **16.25 kg**

CG = **-0.461 cm** = **-0.46 cm**

CG is located 0.46 cm to the left of the datum.

RELOCATION OF WEIGHT ON A BALANCE SCALE

From the calculation shown for fig. 2.7, it is clear that the CG is located at 0.46 cm (0.46 cm to the left of the datum). There is a total mass of 16.25 kg, which has a total moment of 7.5 kg cm.

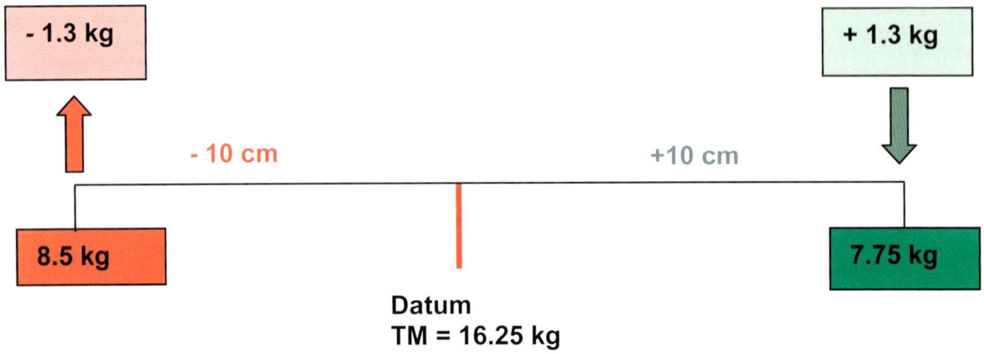

Fig. 2.8

Continuing from this calculation, the effect on the total moment can be calculated and, hence, CG if a 1.3 kg mass is relocated from the left arm location to the right arm location. See fig. 2.8. This is based on the mathematical rule of minus times a minus equals a plus and plus times a plus equals a plus. The movement of this mass has a cumulative effect. This comes from two individual effects, the positive effect of removing a mass from the negative arm and the positive effect of adding the same mass onto the positive arm.

Item	**Mass kg**		**Arm cm**	**- Moment kg cm**	**+ Moments kg cm**
Total mass	16.25	X	-0.46	-7.475	
Left arm	-1.3	X	-10		+13
Right arm	+1.3	X	+10		+13
Totals	**16.25 kg**			-7.475	+26
			Total Moment	+18.525 kg cm	

CG = +18.525 kg cm ÷ **16.25 kg**

CG = +1.14 cm

The CG is now located 1.14 cm to the right of the datum.

The distance of the CG from the datum represents the balance arm for all the masses of the scale. The total mass remains constant, but the total moment changes as the mass is relocated. If the relocation is from the positive arm to the negative arm, the effect is to reduce the positive moment and increase the negative moment. This is based on the rule minus times a plus equals a minus and a plus times a minus equals a minus.

PRACTICE 2

The worked answers are given at the end of this chapter beginning on page 2-16.

Working to 3 pd and giving the answer to 2 pd, calculate the following:

1. The left balance arm of a scale has a weight of 8.5 kg, the right arm has a weight of 7.75 kg, and both arms are 10 cm in length measured from the datum. The datum is located on the pivot point. The CG is located 0.46 cm to the left of the datum. Calculate the new CG if a mass of 3.25 kg is relocated from the right arm to the left arm.

2. The left balance arm has a weight of 208.5 kg, -3.5 m from the datum. The right arm has a weight of 175 kg, +3.5 m from the datum. The datum is located 4.75 m to the right of the pivot point. Working to 3 pd and giving the answer to 2 pd, calculate the CG position in relation to the datum.

3. Working from the answer in question 2 above, calculate the new CG position if 18 kg is relocated from the left arm to the right arm. Work to 3 pd and give answer to 2 pd.

ADDING MASS TO A BALANCE SCALE

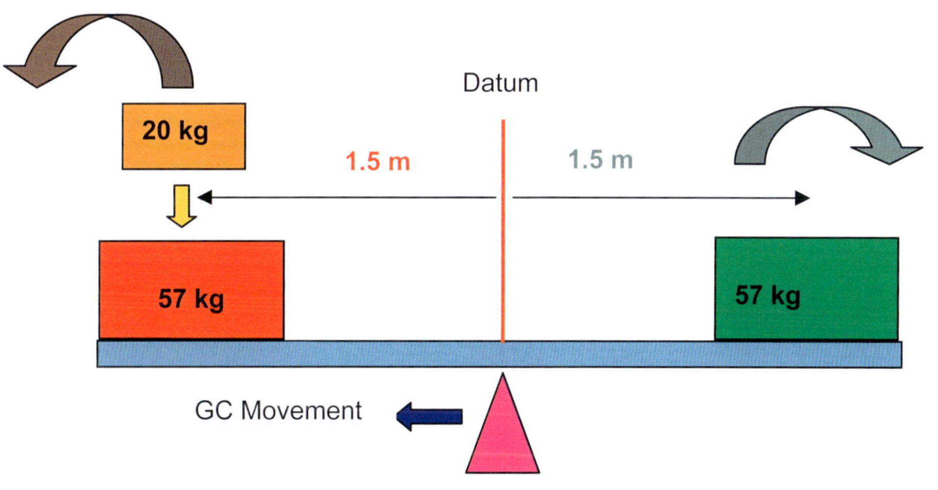

Fig. 2.9

If additional weight is to be added to a balance scale, the overall mass increases and the moment of the arm to which the mass is added increases. This results in the CG moving, as can be seen above in fig. 2.9.

If mass is removed from a balance scale arm, the reverse happens; the overall mass reduces. The moment for the arm from which the mass is removed decreases and the CG moves. In both cases, the CG always moves in the direction of the greater mass. The first calculation below shows the CG on the Datum. The second calculation shows the effect on the CG of adding 20 kg of extra mass to the left arm.

Starting Point

Item	**Mass kg**		**Arm m**	**- Moment kg m**	**+ Moments kg m**
Left arm	57	X	-1.5	-85.5	
Right arm	57	X	+1.5		+85.5
Totals	**114 kg**				
			Total Moment	0.0 kg m	

CG = 0.0 kg m ÷ **114 kg**

The CG is located 0.0 m on the datum.

Addition of 20 kg to left arm

Item	**Mass kg**		**Arm m**	**- Moment kg m**	**+ Moments kg m**
Total mass	114	X	0.0	0.0	0.0
Left arm	20	X	-1.5	-30	
Totals	**134 kg**				
			Total Moment	-30 kg m	

CG = -30 kg m ÷ **134 kg**

CG = -0.224 m

If a mass of 20 kg is added to the left arm, the CG relocates to 0.22 m to the left of the datum.

Using the above calculation as the starting point, the following calculation shows the effect of removing 57 kg from the left arm.

Subtraction of 57 kg from left arm

Item	**Mass kg**		**Arm m**	**- Moment kg m**	**+ Moments kg m**
Total mass	134	X	-0.224	-30	
Left arm	-57	X	-1.5		+85.5
Totals	**77 kg**				
			Total Moment	+55.5 kg m	

CG = +55.5 kg m ÷ **77 kg**

CG = +0.721

= +0.72 m

If a mass of 57 kg is removed from the left arm, the CG relocates to a point 0.72 m to the right of the datum.

PRACTICE 3

The worked answers are given at the end of this chapter beginning on page 2-17.

1. A beam that has equal arms of 3.45 m either side of a central pivot point has a mass of 67 kg placed on its left arm and 37 kg on its right arm. A mass of 16 kg is added to the right arm, and a mass of 11.5 kg is transferred from the left arm to the right arm. Calculate the CG's movement working to 3 pd and giving the answer to 2 pd.

2. From the diagram below, fig. 2.10, calculate the location of the CG in relation to the datum in inches using kg in moments.

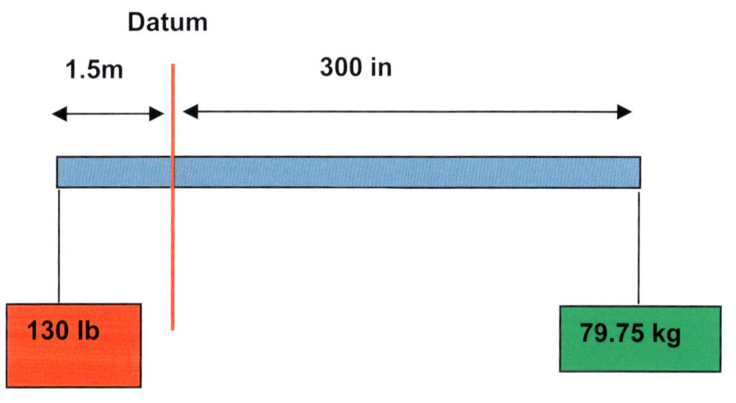

Fig. 2.10

3. The following masses are applied to a beam of 2.5 metres length with the datum located at the left end. Calculate the location of the CG to 2 pd.
   Left arm 13 kg and 14 lb
   Right arm 17 kg and 6.5 lb

BEAM BALANCE

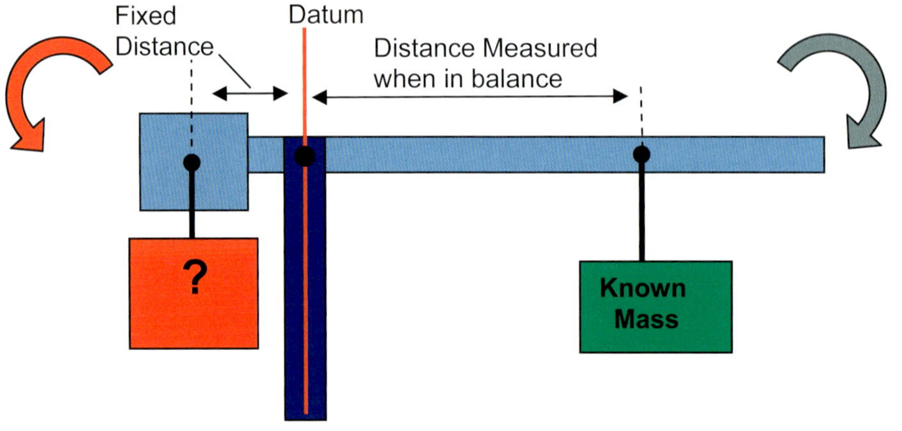

Fig. 2.11

All the preceding examples and practices have been based around a set of scales where the arms are of equal length. To find the mass of any item, an equal mass must be added to the other arm. A method which uses moments to determine the mass of an unknown item is the beam balance or steel yard arm, as shown in fig. 2.11 and fig. 2.12.

Fig. 2.12

If the known mass is 560 lbs (¼ ton), the measured distance is 1095 inches and the fixed distance is 100 inches when the beam is in balance, the sum of the clockwise moments equals the sum of the anticlockwise moments. Therefore, by dividing the anticlockwise moments by the known distance, it is possible to find out the unknown mass of the object (in this case an elephant).

Balance Beam in Balance.

Item	Mass lb		Arm in	- Moments lb in	+ Moments lb in
Right arm	560	X	+1095		+613 200
Left arm	?	X	-100	-613 200	

As Balance Beam in Balance **Total Moment is** **0.0 lb in**

Therefore negative moments must equal **-613 200 lb in**

-613 200 lb in ÷ 100 in = 6132 lb which is approximately 2.5 tons

PRACTICE 4

The worked answers are given at the end of this chapter on page 2-19.

1. Calculate the mass of an object acting over an arm of -3.75 m when it is balanced by a mass of 30 kg acting over an arm of +13.35 m.

2. A mass of 112 lb acting over an arm of 30 ft is used to balance a mass acting over an arm of 4.5 ft. Calculate the balanced mass.

3. An unknown mass is balanced out by 25 kg acting over 8 metres. The unknown mass acts over 1 metre. What is the unknown mass?

DETERMINING THE BALANCE POINT USING A SMALL MASS

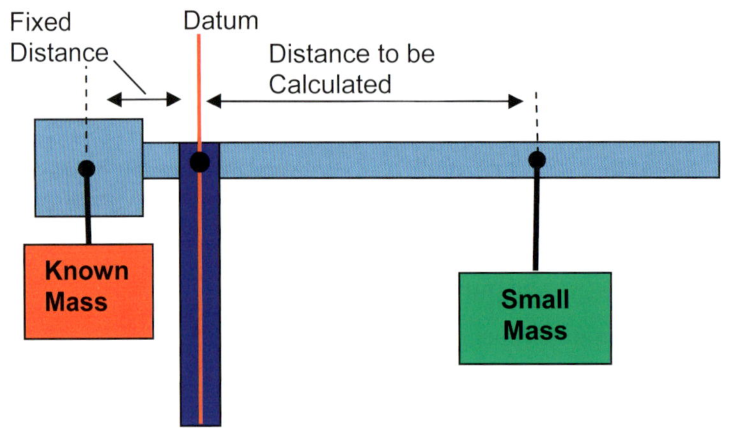

Fig. 2.13

The system also allows us to determine at what point a small known mass would need to be positioned to balance another larger known mass and vice versa. See fig. 2.13.

Example

A mass of 300 kg acting over 2 m is to be balanced by a mass of 25 kg.

Balance Beam in Balance

Item	Mass kg		Arm m	- Moment kg m	+ Moments kg m
Left arm	300	X	-2	-600	
Right arm	25	X	?		+600

As Balance Beam in Balance **Total Moment** is **0.0 kg m**

Therefore Positive moments must equal **+600 kg m**

+600 kg m ÷ 25 kg = 24 m

PRACTICE 5

The worked answers are given at the end of this chapter on page 2-20.

1. A mass of 300 kg acting over an arm of 0.9 m when balanced by a mass of 18 kg would require a balancing arm of?

2. A beam has a mass of 90 kg at an arm of -1.7 m and a balancing mass of 8.5 kg at an arm of +25 m. Give the location of the beam's CG and distance that the balancing mass must be moved to place the beam in balance.

3. What mass is required to balance a 6132 lb object acting over 100 inches if the BA is 7995 inches? See fig 2.14.

Fig. 2.14

ANSWERS

PRACTICE 1

Mass (weight)		Arm	→	Mass kg		Arm m	Moments kg/m
10 lb	X	10 cm	→	4.536	X	0.1	0.454
10 kg	X	10 ft	→	10	X	3.048	30.48
10 lb	X	10 ft	→	4.536	X	3.048	13.826

PRACTICE 2 — QUESTION 1

Item	Mass kg		Arm cm	- Moment kg cm	+ Moments kg cm
Total Mass	16.25	X	-0.46	-7.475	
Left arm	+3.25	X	-10	-32.5	
Right arm	-3.25	X	+10	-32.5	
Totals	16.25 kg			-72.475	

Total Moment -72.475 kg cm

CG = -72.475 kg cm ÷ 16.25 kg

CG = -4.46 cm

The CG is now located 4.46 cm to the left of the datum.

PRACTICE 2 — QUESTION 2

Item	Mass kg		Arm m	- Moment kg m	+ Moments kg m
Left arm	208.5	X	-3.5	-729.75	
Right arm	175	X	+3.5		+612.5
Totals	383.5 kg			-729.75	+612.5

Total Moment -117.25 kg m

CG = -117.25 kg m ÷ 383.5 kg

CG = -0.306 m (3 decimal places)

CG = -0.31 m (2 decimal places)

The CG is now located 0.31 m to the left of the datum.

Note: It is because the position of the CG in relation to the datum is required, the pivot point is irrelevant.

PRACTICE 2 — QUESTION 3

Item	**Mass kg**		**Arm m**	**- Moment kg m**	**+ Moments kg m**
Totals	383.55		-0.306	-117.351	
Left arm	-18	X	-3.5		+63
Right arm	+18	X	+3.5		+63
Totals	383.5			-117.351	+126
			Total Moment	+8.649 kg m	
	CG	=	+8.649 kg m ÷ 383.5 kg		
	CG	=	+0.023 m		
	CG	=	+0.02 m		

The CG is now located 0.02 m to the left of the datum.

PRACTICE 3 — QUESTION 1

Part 1

Item	**Mass kg**		**Arm m**	**- Moment kg m**	**+ Moments kg m**
Left arm	67	X	-3.45	-231.15	
Right arm	37	X	+3.45		+127.65
Totals	104			-231.15	+127.65
			Total Moment	-103.5 kg m	
	CG	=	-103.5 kg m ÷ 104 kg		
	CG	=	-0.995 m		

Part 2

Item	**Mass kg**		**Arm m**	**- Moment kg m**	**+ Moments kg m**
Totals	104		-0.995	-103.5	
Left arm	-11.5	X	-3.45		+39.675
Right arm	+11.5	X	+3.45		+39.675
Right arm	+16		+3.45		+55.2
Totals	120 kg			-103.5	+134.55
			Total Moment	+31.05 kg m	
	CG	=	+31.05 kg m ÷ 120 kg		
	CG	=	+0.259 m		

Distance from original CG to datum	-0.995 m
Distance from new CG to datum	+0.259 m
Distance CG moved	+1.254 m

Therefore, the CG has moved 1.25 m to the right.

PRACTICE 3 — QUESTION 2

Convert 1.5 m to inches = [1.5 ÷ 0.3048] x 12 = 59.055 inches

To convert 1.5 m into inches, divide 1.5 by the constant given on page 4 of the CAP. This gives the number of feet in 1.5 m. This answer multiplied by 12 gives the number of inches.

Convert mass 130 lb into kg = 130 x 0.4536 = 58.968 kg

Item	Mass kg		Arm in	- Moment kg in	+ Moments kg in
Left arm	58.968	X	-59.055	-3482.355	
Right arm	79.75	X	+300		+23 925
Totals	138.718 kg			-3482.355	+23 925
			Total Moment	20 442.645 kg in	
	CG	=	20 442.645 kg in ÷ 138.718 kg		
	CG	=	147.368 in		
	CG	=	147.37 in		

The CG is located 147.37 in to the right of the datum.

PRACTICE 3 — QUESTION 3

Convert mass 14 lb into kg = 14 x 0.4536 = 6.35 kg
Convert mass 6.5 lb into kg = 6.5 x 0.4536 = 2.948 kg

Item	Mass kg		Arm m	- Moment kg m	+ Moments kg m
Left arm	19.35	X	0.0	0.0	0.0
Right arm	19.948	X	+2.5		+49.87
Totals	39.298 kg			0.0	+49.87
			Total Moment	49.87 kg m	
	CG	=	49.87 kg m ÷ 39.298 kg		
	CG	=	1.269 m		
	CG	=	1.27 m		

The CG is located 1.27 m to the right of the datum.

PRACTICE 4 — QUESTION 1

Item	Mass kg		Arm m	- Moment kg m	+ Moments kg m
Left arm	30	X	13.35		+400.5
Right arm	?	X	-3.75	-400.5	
			Total Moment	0	

Therefore, negative moment must equal -400.5 kg m, and the mass must therefore be -400.5 ÷ -3.75 = 106.8 kg

PRACTICE 4 — QUESTION 2

Item	Mass lb		Arm ft	- Moment lb ft	+ Moments lb ft
Left arm	112	X	+30		+3360
Right arm	?	X	-4.5	-3360	
			Total Moment	0	

Therefore, negative moment must equal -3360 lb ft, and the mass must therefore be -3360 ÷ -4.5 = 746.67 lb

PRACTICE 4 — QUESTION 3

Item	Mass kg		Arm m	- Moment kg m	+ Moments kg m
Left arm	25	X	+8		+200
Right arm	?	X	-1	-200	
			Total Moment	0	

Therefore, negative moment must equal -200 kg m, and the mass must therefore be 200 kg.

PRACTICE 5 — QUESTION 1

Item	Mass kg		Arm m	- Moment kg m	+ Moments kg m
Left arm	300	X	-0.9	-270	
Right arm	18	X	?		+270

+270 kg m ÷ 18 kg = 15 m

Balance arm length is 15 m.

PRACTICE 5 — QUESTION 2

Item	Mass kg		Arm m	- Moment kg m	+ Moments kg m
Left arm	90.0	X	-1.7	-153	
Right arm	8.5	X	+25		+212.5
Totals	98.5				
			Total Moment is	59.5 kg m	

Current CG = 59.5 kg m ÷ 98.5 kg = +0.604 m

To balance the beam using the 8.5 kg balance mass, positive moments must equal +153.
Therefore, 153 kg m ÷ 8.5 kg = 18 m.

The balancing mass must be placed at an arm of +18 m, requiring the balancing mass to be moved 7 m to the left.

PRACTICE 5 — QUESTION 3

Item	Mass lb		Arm ft	- Moment lb in	+ Moments lb in
Left arm	6132	X	-100	-613 200	
Right arm	?	X	+7995		+613 200

-613 200 lb in ÷ 7995 in = 76.698 lbs

Balancing mass is 76.7 lbs

Chapter 3
Factors Affecting Mass and Balance in Aircraft

INTRODUCTION

Fig. 3.1

Each aircraft is manufactured using many individual components, each having its own mass and CG. A completed aircraft's mass is the sum of all the components, and its CG location is the sum of all the individual components, acting over their respective arms from the datum.

Basic Empty Mass (Basic Mass) is an aeroplane's mass plus standard items such as:

- Unusable fuel and other unusable fluids
- Lubricating oil in the engine and auxiliary units
- Fire extinguishers
- Pyrotechnics
- Emergency oxygen equipment
- Supplementary electronic equipment

The mass and CG of a newly manufactured aircraft is termed the Basic Empty Mass (BEM) or Basic Mass (BM) and the CG location as BEM CG or BM CG. This is the lightest that a completed aircraft can be.

Adding any other items such as fuel, crew, passengers, and cargo, etc., results in the aircraft getting heavier and the CG moving. This weight increase and CG movement have to be calculated to ensure that the mass and CG remain within the manufacturer's limits. Calculations are carried out prior to any actual loading of the aircraft to ensure that the structure of the aircraft is not damaged and that the CG remains within prescribed limits, both on the ground and in all phases of flight.

To understand how the location of the CG affects flight, it is necessary to understand some of the basic aerodynamic principles.

FORCES IN STEADY LEVEL FLIGHT

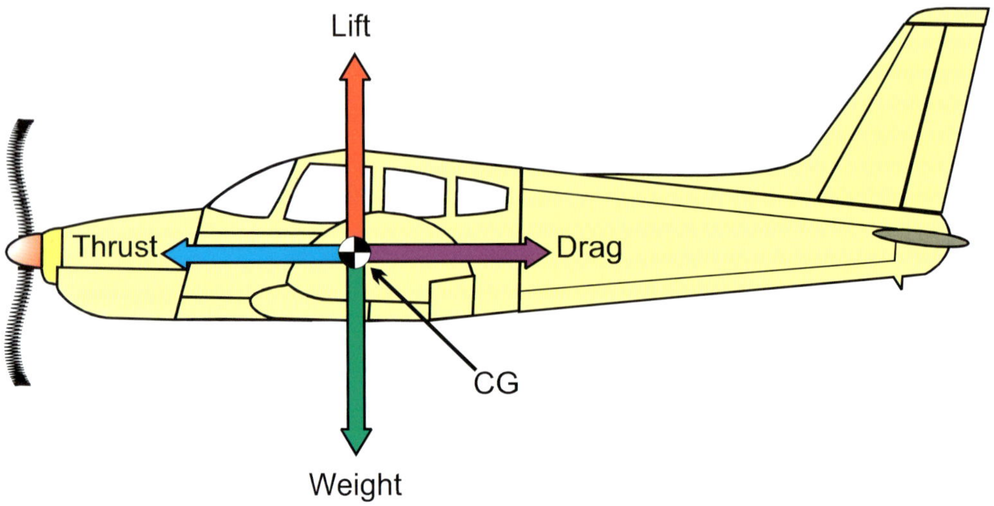

Fig. 3.2

There are four forces acting on an aircraft in flight:

Lift	The force generated by the airflow over the wings
Weight	The mass of the aircaft acted on the acceleration of gravity
Thrust	Developed by the propulsion system
Drag	Developed by the forward speed of the aircraft due to its shape, etc.

These forces form two couples: Lift–Weight and Thrust–Drag.

When an aircraft is in steady, straight-and-level flight, these four forces are said to be in equilibrium or trimmed condition. In this condition, the forces act through the CG of the aircraft, as shown in fig. 3.2.

LIFT

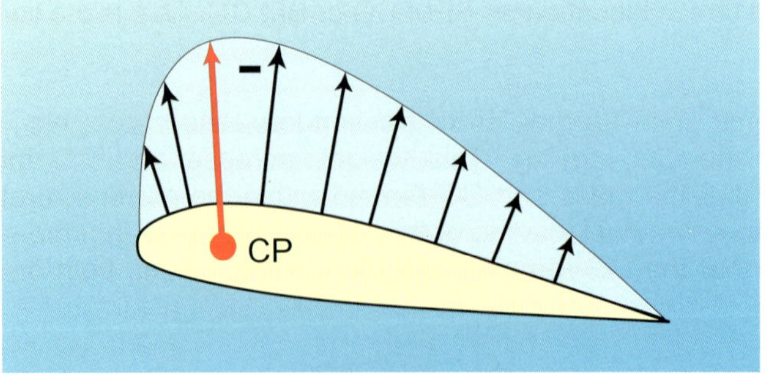

Fig. 3.3

In fig. 3.3, lift is generated due to the drop in the static pressure of the air flowing over the upper surface of the wing. This is due to an increase in velocity as the air flows over the cambered surface. As can be seen, there is a variation of pressure from the leading edge to the trailing edge

TAILPLANE

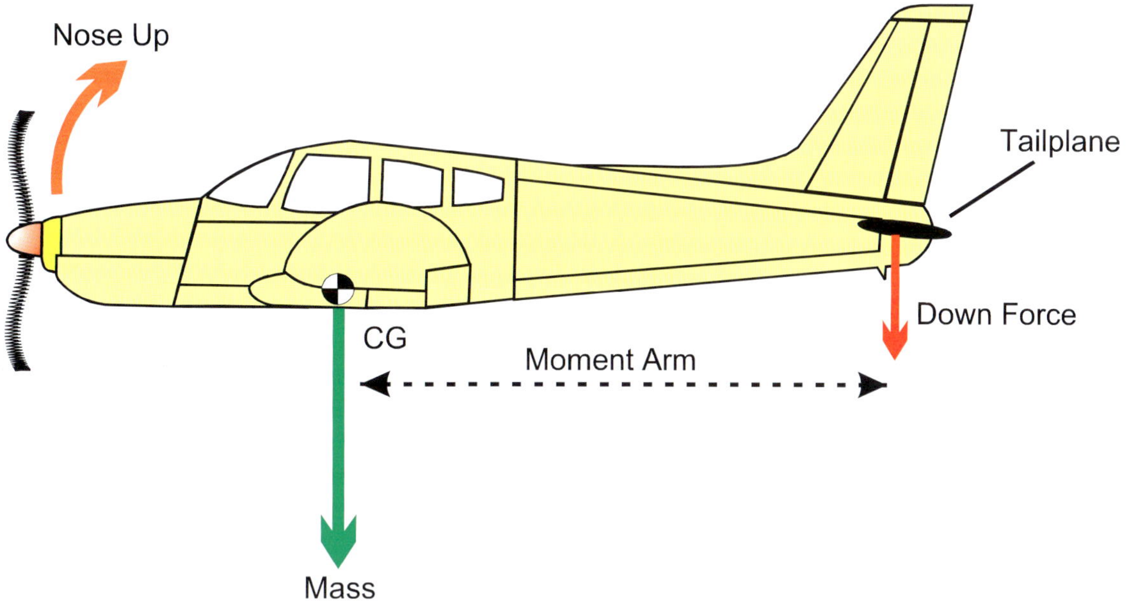

Fig. 3.8

The tailplane is able to supply the force necessary to balance any residual pitching moments. As it is positioned some distance from the aircraft's centre of gravity, this gives it a comparatively large moment arm, allowing a smaller aerodynamic surface to balance out the residual pitching moments created by the larger forces of the two couples acting about the CG.

For the remainder of the Mass and Balance subject, only the Lift and Mass (Weight) couple is considered.

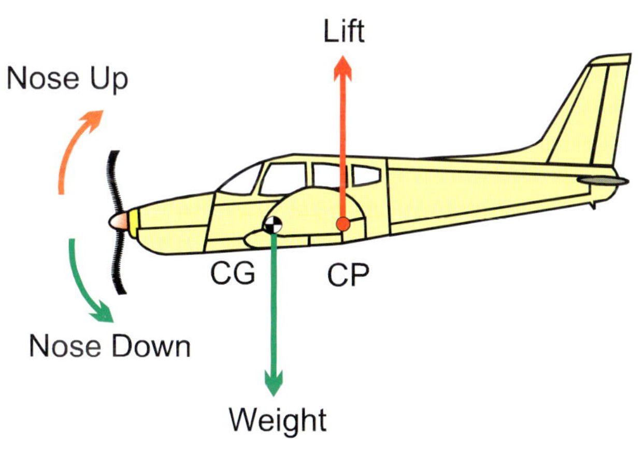

Fig. 3.9

The aerodynamic force created by the tailplane as either a down-force creating a nose-up pitching moment or an up-force creating a nose-down pitching moment is the product of air flowing over the tailplane's surface. An increase in airflow results in an increase in aerodynamic force and vice versa. In some flight conditions (e.g. slow speed), the aerodynamic force produced by the tailplane is insufficient to balance out the pitching moment created by the CP and CG location.

In these cases, the elevator is deflected to alter the camber (curvature) of the tailplane, thus increasing or reducing the aerodynamic forces. The elevators are also deflected to enable the aircraft to climb or dive. Any deflection of the elevators creates drag, which is referred to as Trim Drag. Trim drag increases the aircraft's overall drag and as drag is created through the aircraft's movement through the air, this absorbs engine power in its creation.

STABILITY, CONTROLLABILITY, AND STALLING

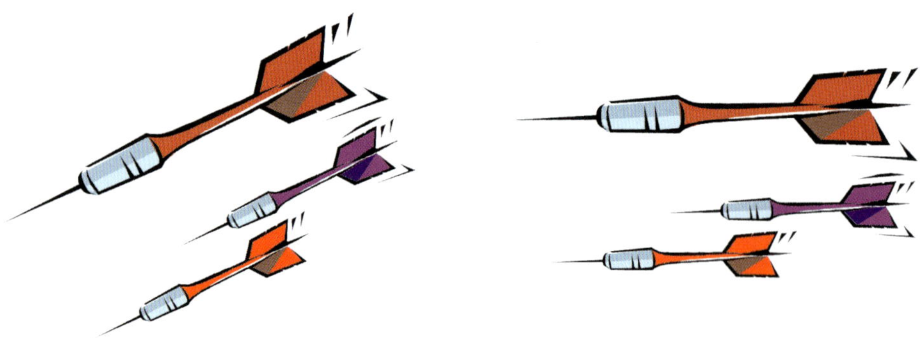

Fig. 3.10

Two of the major considerations a designer has to meet in building an aircraft are stability and controllability. The stability function can be illustrated by darts as shown in the diagram above. The CG of a dart is well forward. It has tail stabilising surfaces at a distance behind the CG, giving the surfaces a lever arm. When the dart is thrown, airflow over its stabilising surfaces produces an up-force and down-force. Should a disturbance in its flight path occur, the stabilising surfaces return the dart to equilibrium; this is stability. However, as the dart's velocity decreases, the stabilising surfaces are unable to maintain level flight, and the tip of the dart drops due to the forward CG and the tail follows; the dart still has stability. If the CG was moved further aft, any disturbance would cause the dart to continue in an erratic manner; this is instability. Between total stability and instability lies controllability.

CG CONSTRAINTS

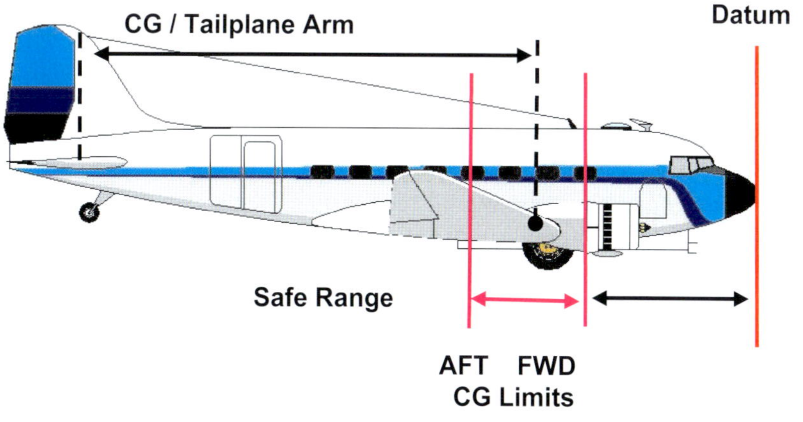

Fig. 3.11

For aircraft, the further forward the CG, the greater the stability; in other words, the less controllable the aircraft is. The further aft the CG, the greater the controllability; the less stable the aircraft is.

To ensure that the aircraft is correctly controllable and stable, the manufacturer places front and rear constraints or limits for the location of the Centre of Gravity, both on the ground and in flight. In fig. 3.11, these are shown as two vertical lines. These lines are referenced from the datum. Provided the aircraft's CG falls on or within these lines, the Centre of Gravity is in limits. The distance between the front and rear CG constraints is termed the safe range, CG range, or normal CG range.

FORWARD LIMIT

The forward limit is determined by the authority that the tailplane has to trim out the increased nose-down pitching moment created by the location of the CG. This authority is determined by the range of the elevator, the airspeed of the aircraft, and the size of the tailplane and the lever arm.

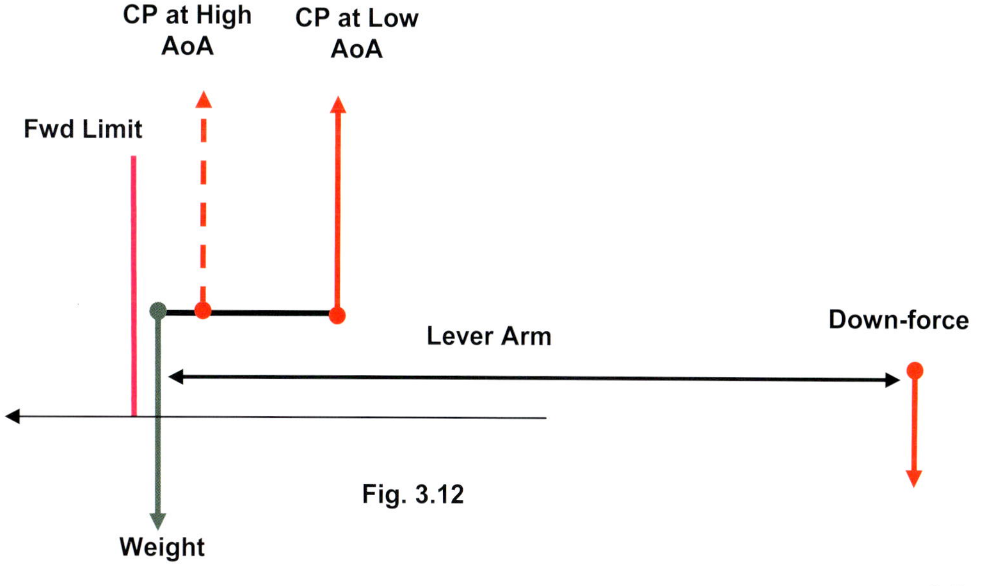

Fig. 3.12

The slowest speed at which an aircraft can safely fly is just above its stalling speed. This increases as the CG moves forward. The aerodynamic reasons for this are outlined below.

Flying at a slower airspeed requires a large angle of attack, which in turn increases the induced drag, increasing the total drag. The increased AoA results in the CP moving forward on the wing, which reduces the nose-down pitching moment created by the mass/lift couple. However, the tailplane's authority is governed by the aerodynamic force it can create; the slow speed reduces the aerodynamic force.

To regain control authority, the pilot has to increase the up deflection of the elevators, which in turn increases the total drag. When the up deflection is equal to the elevators' full range of movement or the total drag has become too large, any slight reduction in speed results in a stall.

As the CG moves forward, the lever arm from the tailplane to the CG increases. This increases the stick force felt by the pilot; aerodynamic load x lever arm. This is termed the stick fixed force and is the airflow's resistance to the elevators' deflection x lever arm.

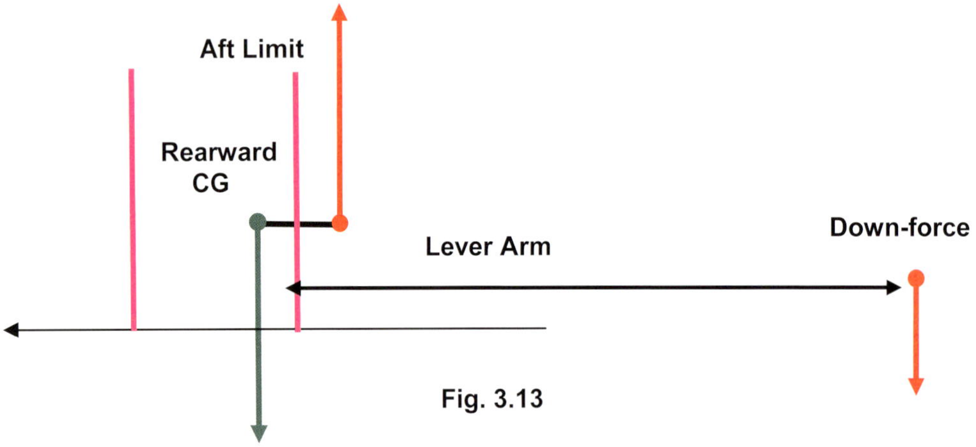

Fig. 3.13

The rear limit is determined by the degree of controllability required. For example, a JAR 25 air transport aircraft requires more stability than a JAR 23 aerobatic aircraft and, therefore, has its rear CG limit further forward in relation to the aerobatic aircraft that is likely to have its fwd limit further aft.

As the CG is located rearward, there is a reduction in the nose-down pitching moment. This reduces down-force requirement, which in turn reduces the need for large elevator deflections (compared with an aircraft of equal mass and fwd CG). This allows for extra control surface deflection, which can be used to alter the pitch attitude of the aircraft. There is also a reduction in trim drag. There is a reduction in the stick fixed force, making it easier for the pilot to move the controls.

NEUTRAL POINT

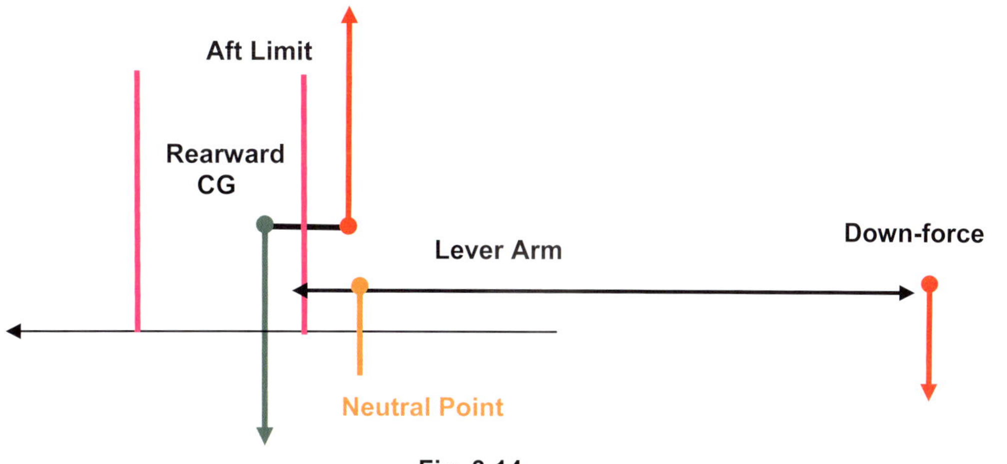

Fig. 3.14

As the CG moves further aft, the controllability increases with the corresponding decrease in stability. The rear limit is set to maintain sufficient stability for the aircraft. The rear limit is set forward of the neutral point. The neutral point occurs on a point termed the aerodynamic centre, which in sub-sonic aircraft is normally located at the ¼ chord point behind the leading edge.

AERODYNAMIC CENTRE

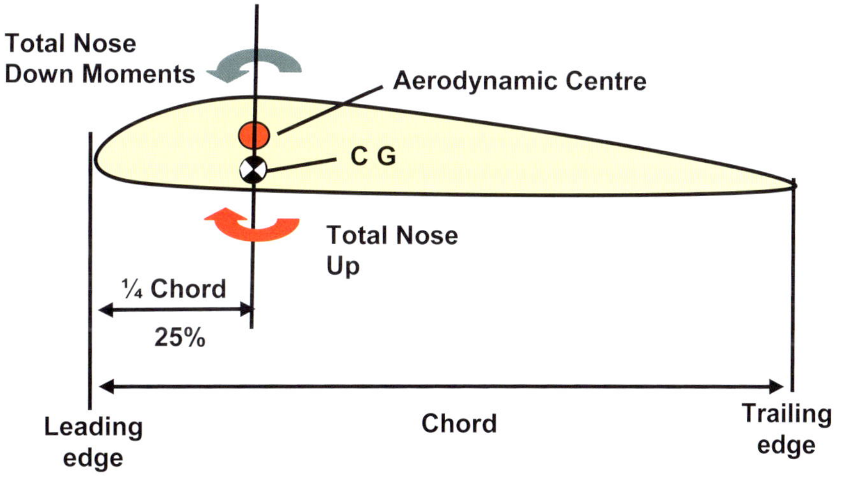

Fig. 3.15

The aerodynamic centre (AC) system is another (more modern) method of reconciling the turning moments created by the forces of Lift, Mass, and Down-force or Up-force. For calculation, the AC is normally considered to be located at 25% chord. In reality, it moves between 23-27% of the chord for a subsonic aircraft. The AC is a very useful point for reconciling the total nose-up pitching moments and the nose-down pitching moments other than at very high angles of attack.

In normal straight-and-level flight, there is a net nose-down pitching moment about the AC. However, if the CG is located on the AC (as per fig. 3.15) there is no net pitching moment. The effect of the tailplane down-force and the lift-mass couple equalling each other at this point is total controllability and neutral stability. This means that if a gust of wind lifts the nose of an aircraft, it remains at the pitch at which the wind leaves it rather than regaining the original pitch. The stick fixed force is nil, so any slight input causes the aircraft to alter its pitch.

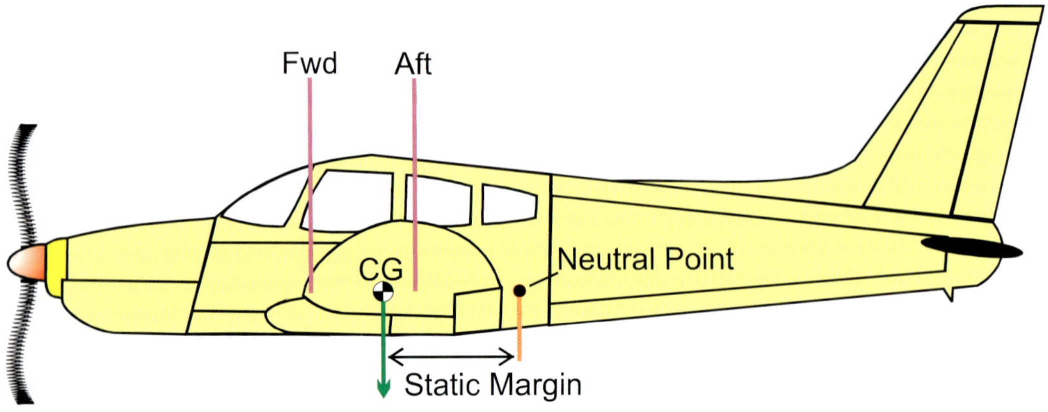

Fig. 3.16

As this is not desirable in an air transport aircraft, the CG rear limit is set forward of this point. This gives the aircraft longitudinal static stability, hence the forward limit. The distance between the neutral point and the CG is termed the static margin or CG margin. Therefore, in real terms, the rear limit is fwd of the neutral point, and the CG is not allowed to get this far aft. If the CG falls aft of the neutral point, the mass of the aircraft acts with the down-force to pitch the aircraft nose up. This reduces the controllability of the aircraft to a point where it is totally uncontrollable.

FACTORS AFFECTING THE LONGITUDINAL CG POSITION IN FLIGHT

Apart from the initial loading of the aircraft, the factors that can affect the longitudinal stability of an aircraft are:

- Fuel consumption
- Flap extension/retraction
- Gear extension/retraction
- Cargo movement

FUEL CONSUMPTION

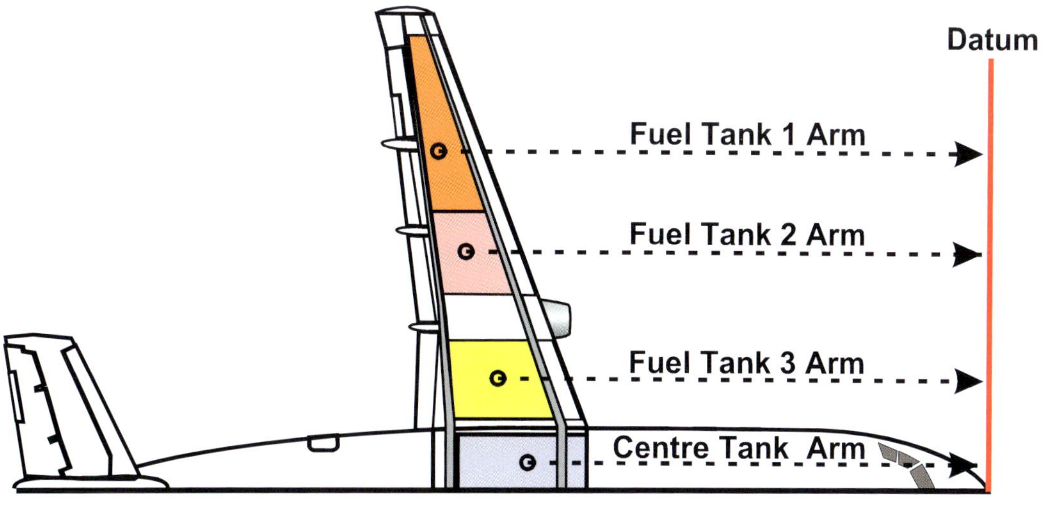

Fig. 3.17

In flight, the aircraft's mass reduces through consumption of fuel. The fuel that is burnt off from the start of the take-off run to landing is known as Trip Fuel. If the aircraft has fuel tanks with varying arms, as the trip fuel is burnt off, the CG position varies due to this consumption and the drop in aircraft mass.

FOWLER FLAPS

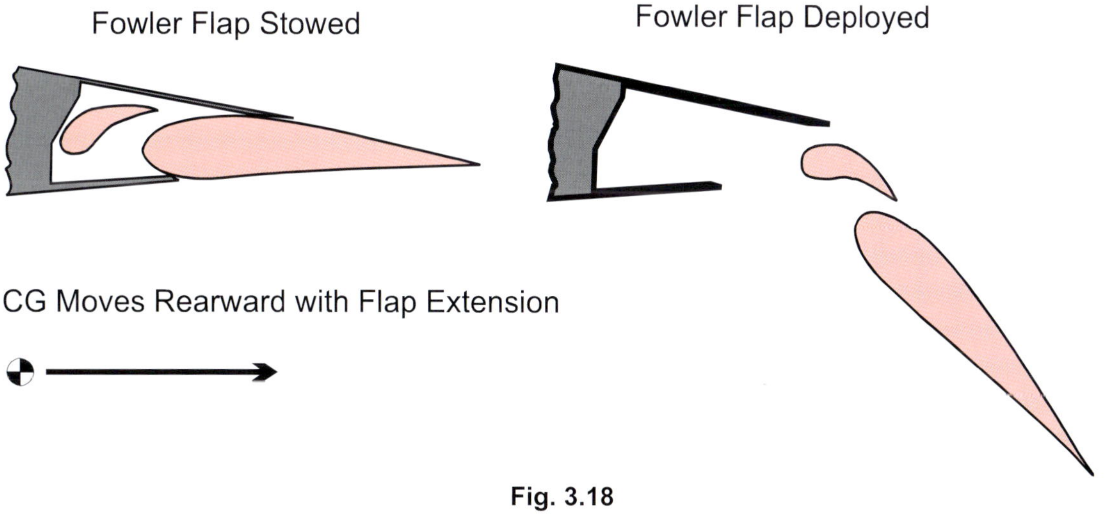

Fig. 3.18

Transport aircraft frequently use Fowler flaps as trailing edge flaps. These translate rearward on extension, moving rearward as well as lowering the trailing edge. This results in the CG moving rearward with the flaps' extension and forward with the flaps' retraction.

LANDING GEAR DESIGN

Fig. 3.19

Depending on the design of the aircraft and its undercarriage, the raising and lowering of the landing gear can have an effect on the CG. Most aircraft have main gears that retract laterally; the movement of these gears has no effect on the longitudinal CG. Under normal circumstances, nose gears retract longitudinally. Their operation is considered to have only a negligible effect. However, if considered in isolation, the raising of a forward retracting nose gear moves the CG forward and vice versa.

High-wing turboprops can have their main gears mounted in the engine nacelles. The retraction or extension of these gears has a significant effect on the CG. If the gears retract forward, the CG moves forward and vice versa.

CARGO

When an aircraft is loaded, the CG and mass must be within the prescribed limits of the aeroplane. If the cargo should shift in flight, the aircraft can become either too stable or uncontrollable, at best leading to extreme difficultly in handling for the crew, at worst a crash. Additionally, cabin crew and passenger movements have an effect on the trim of the aeroplane.

THREE CG POINTS THAT MUST BE CALCULATED

Provided the aircraft is not overloaded, the CG may fall on or between the CG limits. However, as the aircraft's CG location moves with the consumption of fuel, prior to any flight, the aircraft's calculated Take-Off Mass (TOM), its estimated Landing Mass (LM), and Zero Fuel Mass (ZFM) must be compared to the limits. The ZFM is compared to ensure that the CG is in limits when the aircraft is on the ground and that the aircraft would still be able to fly should for some reason the entire fuel load be used or lost.

SUMMARY

As the CG moves toward the fwd limit:

- Stability increases and controllability decreases.
- On take-off, the nose is heavy requiring more elevator deflection to rotate the aircraft. This can result in a later lift-off.
- The climb is suppressed, as the nose wants to pitch down.
- The nose-down pitching moment increases, requiring greater deflection of the elevators, which results in an increase in trim drag.
- The down-force required increases.
- As the CG moves further forward, the down-force increases the wing loading (effectively making the aircraft heavier), which results in a higher stalling speed.
- The increase in trim drag and wing loading requires more power to maintain a given airspeed.
- The increase in power results in more fuel being burned, thus reducing both range and endurance.
- On final approach, the airspeed must be kept up to prevent nosing in, resulting in a higher landing speed.
- In the event of a landing climb, the aircraft's climb is suppressed.

As the CG moves toward the aft limit:

- Stability decreases and controllability increases.
- On take-off, the nose is light, requiring less elevator deflection to rotate the aircraft. This can result in the aircraft rotating early, leading to an early lift-off at a slower than planned speed, which in turn reduces the acceleration of the aircraft.
- The climb angle is increased, as the nose wants to pitch up.
- The nose-up pitching moment increases, requiring greater deflection of the elevators, which results in an increase in trim drag.
- The down-force required reduces. As the CG moves further aft, the down-force decreases the wing loading (effectively making the aircraft lighter) which results in a lower stalling speed.
- The increase in trim drag requires more power to maintain a given airspeed.
- The increase in power results in more fuel being burnt, thus reducing both range and endurance.
- On final approach, the airspeed must be kept up to prevent the tail from dropping. This can result in a higher landing speed.
- In the event of a landing climb, the aircraft's nose rotates more easily but can over pitch and stall.
- Spin recovery becomes difficult because of the possibility of flat spins developing.
- Any glide angle is difficult to maintain because of the aircraft's tendency to pitch up.

If an aircraft's MTOM is exceeded, but the CG is located within the safe range, it results in:

- Having a greater take-off run, because it requires a greater speed to produce the lift required.
- Reduced climb performance as there is less excess power available to be converted into altitude.
- A reduction in airspeed for a given power setting.
- Higher stalling speed. Any reduction in airspeed results in a significant decrease in lift, therefore, the aircraft is more likely to stall.
- The aircraft having a reduced service ceiling due to reduced excess power.
- A higher power setting is required to maintain a given airspeed. This results in an increase in fuel consumption.
- A decrease in both range and endurance, resulting from the higher fuel consumption.
- A higher landing speed to prevent the aircraft from stalling during the approach and landing. This results in a longer landing run.
- Heavy braking on landing to stop the aircraft from exceeding the length of the runway. This can lead to damage to both tyres and brake units.

Fig. 3.20

The greater the condition, the more radical the effects become.

These limits can be given in one or both of two ways, numerically or graphically. The numerical limits are given in the weighing report, as per the limits in the SEP 1 and MEP 1 data sheets pages 5/6 and 12/13 respectively of CAP 696. The graphical format is termed a CG envelope.

CENTRE OF GRAVITY ENVELOPE

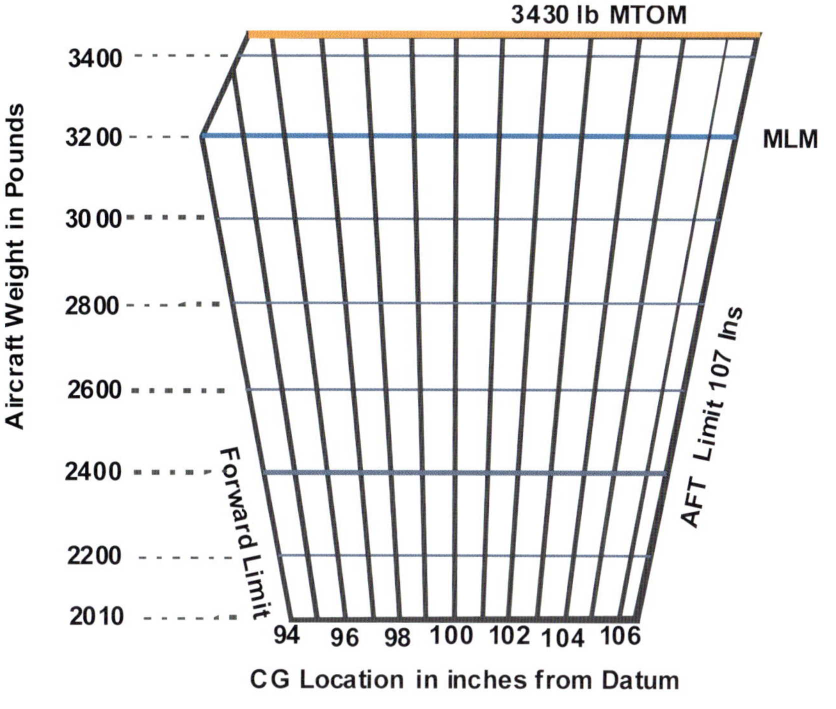

The vertical and inclined lines of the envelope represent the front and rear CG limits. The upper horizontal line represents the maximum take-off mass. Between the base line and the MTOM line, the envelope is marked off with further horizontal lines at set mass intervals. Dependent on the style of CG envelope, the vertical lines are marked off as CG linear positions or in larger aircraft as percentages of the Mean Aerodynamic Chord (MAC).

PLOTTING ON A CG ENVELOPE

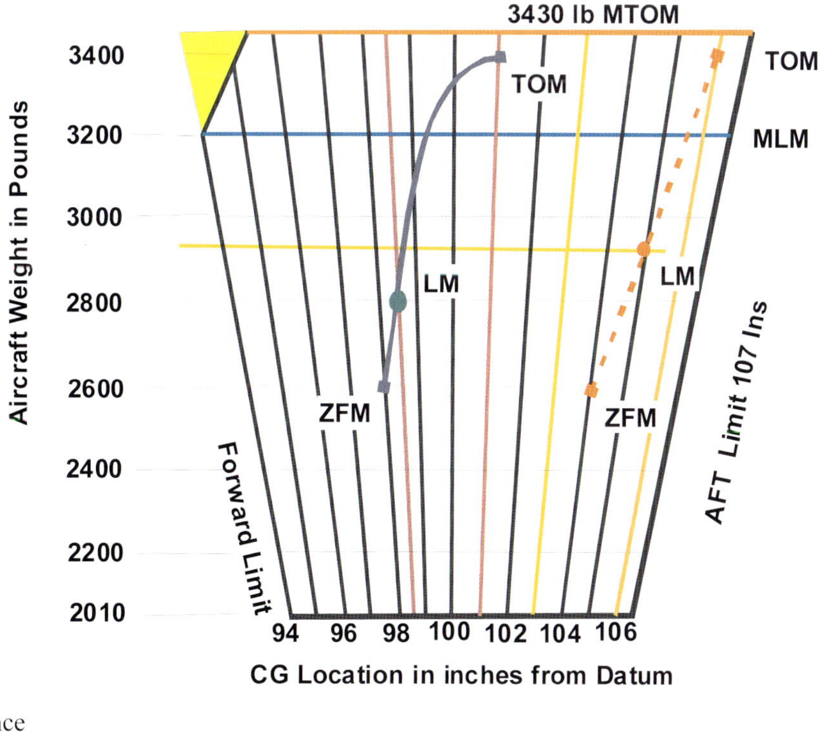

The points to plot are the TOM, LM, and ZFM. If the fuel tanks share a common arm, a straight line drawn from the TOM to ZFM should pass through the LM plot. Refer to the figure above, right-hand plot. If the aircraft has varying fuel tank arms, the line from the TOM to the ZFM shows a curve. Refer to figure above, left-hand plot.

Plotting the TOM, LM, and ZFM plus associated CGs shows whether the intended masses exceed the structural or CG limitation. Plotting the ZFM also shows the aircraft's CG location should for some reason all the fuel be consumed or lost.

The forward CG limit is cut back in this example at 3200 pounds from + 94 in to + 95.5 in at the MTOM of 3430 pounds. The segment that has been removed is coloured yellow to highlight the area. The CG must not fall into this area as the greater mass acting over the arm would induce too much stability.

The CG envelope for some aircraft, such as the SEP 1 of the Loading Manual (found on page 9 of the CAP 696 as fig. 2.5 and shown overleaf), has a series of diagonal lines as well as the inclined verticals and horizontals. The verticals represent the CG position in inches aft of the datum, the horizontals represent mass values, and the diagonals represent moments indexed by 100. This allows the CG to be plotted without making the final CG calculation of Tmo ÷ TM = CG. See example on following page.

Referring to the CG envelope below, a moment of 2700 plotted against a mass of 3400 pounds places the CG at 79.25 inches aft of the datum. As the vertical lines are inclined, they diverge so care must be taken when measuring the scale before plotting the points.

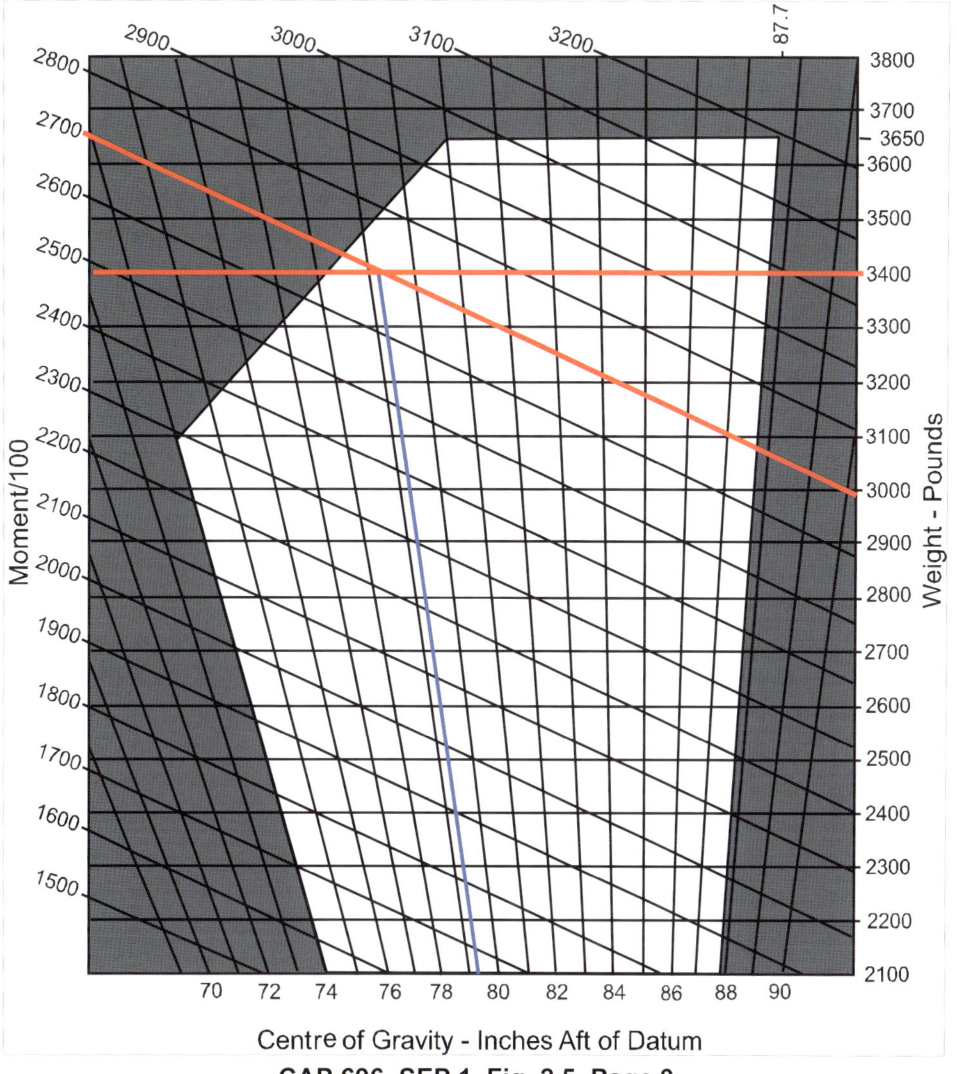

CAP 696, SEP 1, Fig. 2.5, Page 9

Though not shown in the CG envelopes of the CAP 696, some JAR 23 aircraft have areas marked off within the envelope that denote the aircraft's maximum utility mass and the CG limits for this type of flying. Thus, an aircraft can take-off in the normal category and, after sufficient fuel has been burnt, execute utility manoeuvres.

Loading Index (LI) — Is a non-dimensional figure that is a scaled down value of a moment. It is used to simplify mass and balance calculations.

As moments can become very unwieldy in calculations, they can be divided by a constant, making them more manageable. For example, a moment of 12 300 000 kg m could be divided by 100 000, giving a moment of 123 kg m. This value is referred to as an LI.

The LI 123 kg m when multiplied by the indices of 100 000 reverts to the full moment number. If the original moment had been 12 376 987 kg m, when converted into an LI using the same index value of 100 000, would return an LI of 123.76987. As this is again unwieldy, it would be reduced to a given number of decimal places (in this case one), returning an LI of 123.8 kg m. Obviously, this does not return to the original value, but for large aircraft, the values are such that this is not considered a problem. For light aircraft, small constants are used (refer to CAP 696 page 5). Loading indices can be used for all types of aircraft and are used in each of the aircraft types in CAP 696.

Chapter 4
Mass Definitions and Limitations

INTRODUCTION

When an aeroplane has been manufactured and all the basic equipment required for its role (e.g. passenger seats, galley, toilets, etc.) are fitted, the aircraft has to be weighed prior to it entering service and at set intervals throughout its service life. The frequency of these subsequent mass evaluations is covered later.

The initial weighing is to find the aeroplane's **basic empty mass (BEM)**. This includes the:

1. Unusable fuel and other unusable fluids
2. Lubricating oil in the engine and auxiliary units
3. Fire extinguishers
4. Pyrotechnics
5. Emergency oxygen equipment
6. Supplementary electronic equipment

1. In the case of fuel, the term **unusable** means the fuel that cannot be drawn from the tanks to operate the engine. The term **other unusable fluids** covers hydraulic fluid and cooling fluid. These must be charged to their correct level. This does not include the potable water (drinking water) or the lavatory pre-charge.
2. The engines and auxiliary units must be filled to their correct levels.
3. The fire extinguishers (hand-held) must be fitted at the designated location; all fire extinguishers have to be correctly charged.
4. Any pyrotechnics required to be carried or fitted as standard are correctly located.
5. The emergency oxygen equipment is fitted.
6. Any supplementary electronic equipment that is carried as standard is correctly fitted in the given location.

In accordance with JAR-Ops 1, all weighing operations must be carried out in an enclosed building.

The BEM is often used as the starting point for light aircraft mass and balance calculations. However, whilst the BEM will have been calculated for JAR 25 aircraft, the starting point for all operational mass and balance calculations is the Dry Operating Mass (DOM).

Dry Operating Mass (DOM) — Is the total mass of the aeroplane ready for a specific type of operation excluding all usable fuel and traffic load. The mass includes items such as:

1. Crew and crew baggage
2. Catering and removable passenger service equipment
3. Potable water and lavatory chemicals
4. Food and beverages

1. The crew includes the rear crew for passenger operation.
2. The passenger service and catering equipment can vary depending on the role and passenger configuration (e.g. V.I.P. to charter).
3. Potable water is chemically treated (purified) water.
4. Food and beverage vary dependent on role and length of flight.

Note: This definition excludes usable fuel and traffic load. The items included are often termed **variable load**. While this is not a JAR term, or given in the CAP 696, it is in common usage in the aviation industry.

BEM + Variable Load = Dry Operating Mass

Traffic Load — The total mass of passengers, baggage, and cargo, including any non-revenue load.

The **non-revenue load** includes items that the aircraft carries that are not used in flight or are part of the aircraft's role equipment and no financial charge is made. Frequently, air transport aircraft carry spare wheels and brake units when making trips to airports where they do not have these items as standard spares. On other occasions, the aircraft may require carrying ballast to balance out an item of freight.

Note: Traffic load does **not** include usable fuel.

Two further traffic load related definitions are normally regarded as Air Transport Aircraft (JAR 25) terms but can be used for lighter aircraft. They are:

Allowed Traffic Load — The maximum mass of traffic load that an aircraft can carry.

Many aircraft cannot carry both their maximum fuel load and maximum passenger load at the same time due to structural limitation or performance limitations; therefore, a maximum limit is set.

Under Load — This is the difference between the actual traffic load and the allowed traffic load.

This is the term given to the extra mass that the aircraft can carry if it is loaded to the allowed traffic load. For example, if an aircraft has an allowed traffic load of 4000 kg but is only carrying a traffic load of 3500 kg, there is an under load of 500 kg.

Pay Load — This is defined as that part of the traffic load from which the revenue is earned.

At this point, an aircraft at DOM could be loaded in one of two ways. Both conditions have their own definitions, which are given below. However, to put them in the correct context, the fuel definitions need to be understood.

FUEL DEFINITIONS

These are:

1. Block Fuel, also termed Bulk Fuel or Ramp Fuel
2. Start, Run-up, and Taxi Fuel
3. Take-Off Fuel
4. Trip Fuel
5. Landing Fuel
   a. Contingency Fuel
   b. Reserve Fuel

1. **Block or Ramp Fuel** is the full fuel load put into the aircraft. This can take the aircraft above its maximum take-off mass.
2. **Start, Run-up, and Taxi Fuel**, often referred to as start fuel, is an allowance made for consumption of fuel in starting the engines, running them up, and taxiing to the take-off point. This is normally given by airport and depends on the type of aircraft, its stand location, and the runway in use.
3. **Take-Off Fuel (TOF)** is the full load on board the aircraft at the start of the take-off run. Block Fuel less Start Fuel equals Take-Off Fuel.
4. **Trip Fuel (TF)** is the fuel that is estimated to be burnt during the flight from the commencement of the take-off run to the completion of the landing run. In reality, this can vary due to unexpected headwinds or tailwinds, etc.
5. **Landing Fuel** is the estimated fuel load on touchdown and contains the **contingency** and **reserve** fuel elements. Take-Off Fuel less Trip Fuel equals Landing Fuel.

To summarise

Landing Fuel + Trip Fuel	=	Take-Off Fuel
Take-Off Fuel + Start Fuel	=	Ramp Fuel

Operating Mass (OM) — Is the DOM plus fuel but without traffic load.

From DOM, the aircraft can be brought up to OM by loading the Take-Off Fuel (TOF) but none of the passengers, baggage, ballast, and cargo.

DOM + TOF = OM
BEM + VL + TOF = OM

Zero Fuel Mass (ZFM) — Is DOM plus traffic load but excluding fuel.

From DOM, the aircraft can be brought up to ZFM by loading all of the passengers, baggage, ballast, and cargo but none of the fuel.

DOM + Traffic Load = ZFM
BEM + VL + Traffic Load = ZFM

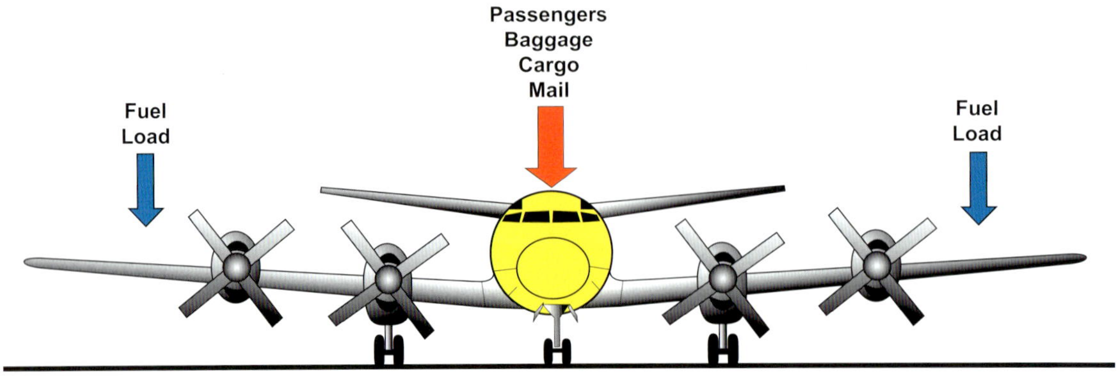

Maximum Zero Fuel Mass

For the majority of aircraft, the fuel is stored in fuel tanks located in the wings outboard of the main wheels, while the traffic load is located in the fuselage. For larger aircraft, the mass of the fuel is used to balance the combined masses within the fuselage, as represented in fig. 4.2.

STRUCTURAL LIMITATIONS

Four structural limitations apply to aircraft. These are:

1. Maximum Structural Taxi Mass
2. Maximum Structural Take-Off Mass
3. Maximum Structural Landing Mass
4. Maximum Zero Fuel Mass

Light aircraft may only use the Maximum Structural Take-Off Mass limitation due to their lower masses and construction.

Maximum Structural Taxi Mass (MSTM) — Is the structural limitation on the mass of the aeroplane at the commencement of taxi.

This allows for the fuel that the aircraft consumes during engine start, run-up, and taxi. This is also referred to as **Maximum Structural Ramp Mass** and can be shortened to **Maximum Taxi Mass** or **Maximum Ramp Mass**.

Maximum Structural Take-Off Mass — The maximum permissible total aeroplane mass at the start of the take-off run.

This is the heaviest mass at which an aircraft can start the take-off run given the most favourable conditions anywhere in the world and is frequently refered to as **Maximum Take-Off Mass (MTOM)**. For an aircraft to actually start the take-off run at this mass, it must have a Maximum Ramp Mass above the MTOM that is equal to or greater than the start fuel requirement. If the MTOM limit is exceeded, the aircraft can suffer structural damage and the Take-Off distance required increases.

Maximum Structural Landing Mass — The maximum permissible total aeroplane mass on landing under normal circumstances.

This is the heaviest mass at which an aircraft can start touchdown given the most favourable conditions anywhere in the world and is frequently referred to as **Maximum Landing Mass (MLM)**. If the MLM limit is exceeded, the aircraft can suffer structural damage and the landing distance increases.

Maximum Zero Fuel Mass (MZFM) — The maximum permissible mass of an aeroplane with no usable fuel.

Because the mass of the fuel in the aircraft's wing tanks is used to balance the internal load, a limit is placed on how much weight may be put into the aircraft's fuselage before fuel has to be added into the wing tanks. If the MZFM limit is exceeded, the aircraft can suffer serious structural damage.

PERFORMANCE LIMITATIONS

The limitations given in the Structural Limitations above define the maximum masses given the most favourable conditions anywhere in the world. The other factors that must be taken into account are performance related. Some of these are:

- The altitude of the airfield } Density
- The air temperature } Density
- The length of the runway
- The topography of the area

This is not the complete list. Be aware that to meet the performance requirements, an aircraft's mass can be limited. These limits have to be taken into account not only for take-off, but also for landing. In some cases, the conditions enroute may also become limiting factors such as flying over mountainous terrain.

Performance Limited Take-Off Mass (PLTOM) — Is the take-off mass subject to departure airfield limitations.

Performance Limited Landing Mass (PLLM) — Is the mass subject to the destination airfield limitations.

Where a performance limitation is given for an airfield, it can exceed the structural limitation of a particular aircraft (e.g. a performance limitation for a B747 heavy passenger jet is not necessarily a performance limitation for a BAe 146 medium pasenger jet. In any case, where a performance limit is given, it must be compared with the structural limitation for the aeroplane in question and the lower of the two used as the limiting factor. This is known as the **Regulating Mass**.

Regulated Take-Off Mass (RTOM) — Is the lowest of "Performance Limited" and "Structural Limited" Take-Off Mass.

The RTOM can also be referred to as **Maximum Allowable Take-Off Mass (MATOM)**.

Regulated Landing Mass (RLM) — Is the lowest of "Performance Limited" and "Structural Limited" Landing Mass.

The RLM can also be referred to as **Maximum Allowable Landing Mass (MALM)**.

Take-Off Mass (TOM) — Is the mass of the aeroplane including everything and everyone contained within it at start of the take-off run.

The actual mass of the aeroplane at take-off is called the Take-Off Mass (TOM). This is the regulated mass.

Landing Mass (LM) — Is the mass of the aeroplane including everything and everyone contained within it at start of the landing run.

The actual mass of the aeroplane at landing is called the Landing Mass (LM). This is the regulated mass.

Gross Mass — The mass at the given condition.

This term can be used in JAR examinations and the correct answer reflects the conditions given in the question.

Useful Load — The mass of usable fuel and traffic load.

This term has been defined in the JAR learning objectives and is used in JAR examinations. The correct answer reflects the definition above.

Datum or Reference Datum — (Relative to an aeroplane) is that (vertical) plane from which the centres of gravity of all masses are referenced.

The datum in aircraft is considered to be a vertical or perpendicular line or plane. The location of an aircraft's datum is decided by the manufacturer and can be anywhere within the fuselage, in front of it, or behind it. Wherever the manufacturer decides to locate the datum for the aircraft, it is the point from which all balance arms are measured.

Each component that is used to make up an aeroplane or item that is loaded into the aeroplane has its own mass and centre of gravity through which its mass is said to act. All these separate masses have balance arms, their distance from the datum.

The locations of the datum for each of the three aircraft of CAP 696 are:

SEP 1 Page 5

In the diagram of the light single-engine aircraft, the Firewall is first located to locate the Datum:

Reference Datum 39.00 inches forward of the firewall

As the location of the datum for this aeroplane is not a physical item but an arbitrary position decided by the manufacturer, the firewall is used as a reference point from which to take a measurement to locate the datum. All balance arm measurements are made from the reference datum.

MEP 1 Page 12

In the light twin-engine aircraft, the reference datum is found 78.4 inches forward of the reference point. This is the leading edge of the wing inboard of the inboard edge of the inboard fuel tank.

MRJT 1 Page 20

In this diagram, the datum is located in the nose section of the aircraft, 540 inches forward of the Front Spar (FS). The front spar is the reference point.

PRACTICE

1. The traffic load, basic mass, and variable load can be defined as:
   a. DOM
   b. OM
   c. GM
   d. ZFM

2. RLM is:
   a. Always equal or greater than MLM
   b. Always equal or less than MLM
   c. Always equal to MLM
   d. Never equal to MLM

3. MZFM is defined as the:
   a. Minimum permissible mass of an aeroplane with no useable fuel, but including payload
   b. Maximum permissible mass of an aeroplane with useable fuel, but no payload
   c. Minimum permissible mass of an aeroplane with the useable fuel but no payload
   d. Maximum permissible mass of an aeroplane without the useable fuel but including payload

4. The unusable fuel is accounted for as part of:
   a. DOM
   b. Block fuel
   c. BM
   d. RM

ANSWERS

1. d
2. b
3. d
4. c

Chapter 5
Aircraft Weighing and Floor Loading

INTRODUCTION

Fig. 5.1

Aircraft must be weighed periodically throughout their service life. The periodicity of this is covered in Chapter 8 of this volume — *JAR-Ops 1 Requirements*. This section outlines how the actual weighing is carried out. It also explains the basic principle of loading an aircraft's hold.

WEIGHING

The aircraft is set up as required by the manufacturer (see chapter 4) to establish the BEM. There are three systems used to weigh aircraft:

- Hydrostatic units
- Weighbridge
- Electronic systems

Hydrostatic units are very heavy-duty hydraulic pressure gauges that are calibrated in units of mass. They are fitted between aircraft lifting jacks and the strong points on the aircraft's structure, termed jacking points. As the aircraft is lifted clear of the ground, the total weight of the aircraft is felt as a hydraulic pressure in each of the units.

Weighbridges are normally built into the floor of a building and the aircraft is towed onto the weighbridge, which records the weight of each undercarriage leg.

Electronic systems use strain gauges or piezoelectric devices. These electronic units can either take the form of thin plates that are placed under the aircraft's tyres or be fitted between the lifting jacks and the jacking points. The more commonly used systems are those onto which the aircraft is rolled.

Whichever system is used, the mass registered by each unit is termed a **reaction** from the Newtonian law that each and every action has a reaction. Where a reaction is taken from under a wheel, it can also be termed a **wheel mass** or **wheel weight**.

Aircraft must be placed in a level for flight condition to be weighed. To achieve this, it may be necessary to place ballast within the fuselage or fit gags to the undercarriage legs. The weighing units also feel the mass of these items. The total mass of the aircraft is the sum of all the reactions less the mass of ballast, etc.

CALCULATING THE BEM CG

As the distance between the reaction points and the datum are precisely known or can be measured, each reaction point has an arm from the datum, allowing the moments acting about the datum to be calculated.

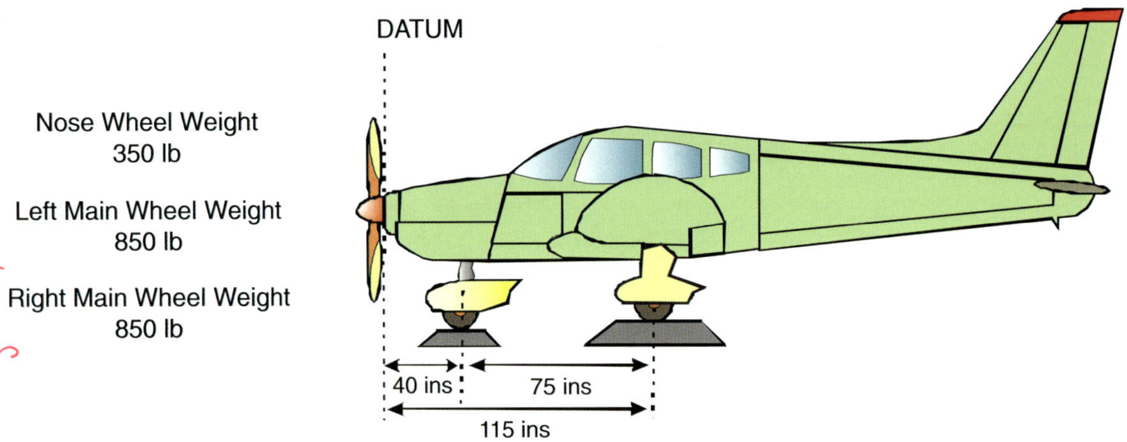

Fig. 5.2

For example, refer to the figure above to find this aircraft's BEM and BEM CG.

Item	Mass - lb		Arm - in	+ Moments lb in	- Moments lb in
Nose wheel	350	X	40	14 000	
Left main	850	X	115	97 750	
Right main	850	X	115	97 750	
Total	2050			195 500	
			Total moment	195 500	

195 500 lb in ÷ 2050 lb = + 95.37 in

Again, if any ballast or gags are located in the aircraft, they contribute to the moments being registered and must be deducted from the calculation. If, as in the calculation below, an aeroplane requires ballast or any other equipment to be fitted to enable it to be weighed, these masses and the associated moments must be deducted.

Item	Mass - kg		Arm - in	+ Moments lb in	- Moments lb in
Nose wheel	1350	X	-40		-54 000
Nose gag	**-13**	X	-40	**+ 520**	
Left main	3850	X	+45	173 250	
Right main	3900	X	+45	175 500	
Ballast	**-200**	X	+30		**- 6000**
Total	8900			349 270	-60 000
			Total moment	+289 270	

+ 289 270 kg in ÷ 8900 kg = + 32.50 in

PRACTICE 1

1. When weighed, the following single wheel reaction masses are recorded:
   Nose wheel 1000 kg, at an arm of - 92.5 in
   Main wheel 3750 kg at an arm of + 68.75 in
   The aircraft has twin nose and main wheels and requires 150 kg of ballast acting at + 350 in.
   Calculate the CG and BEM.

2. From the diagram below, calculate the mass of the aircraft.

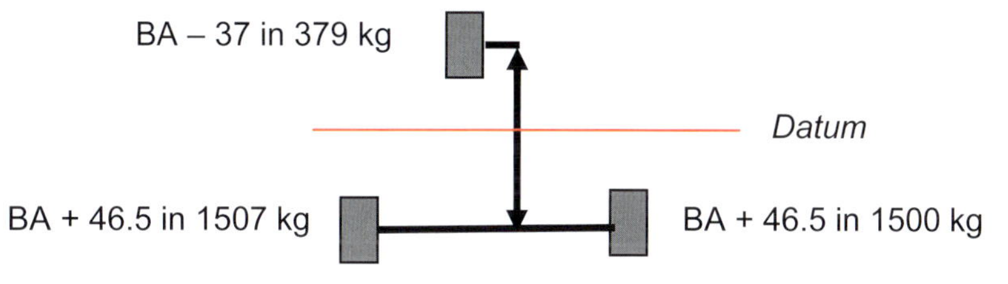

Fig. 5.3

3. Calculate the CG for the following aircraft.

Item	Reaction	Arm	+ Moment	- Moment
Nose Jack	16 677 N	+25 in		
Left Main Jack	34 335 N	+303 in		
Right Main Jack	34 335 N	+303 in		

FLOOR LOADING

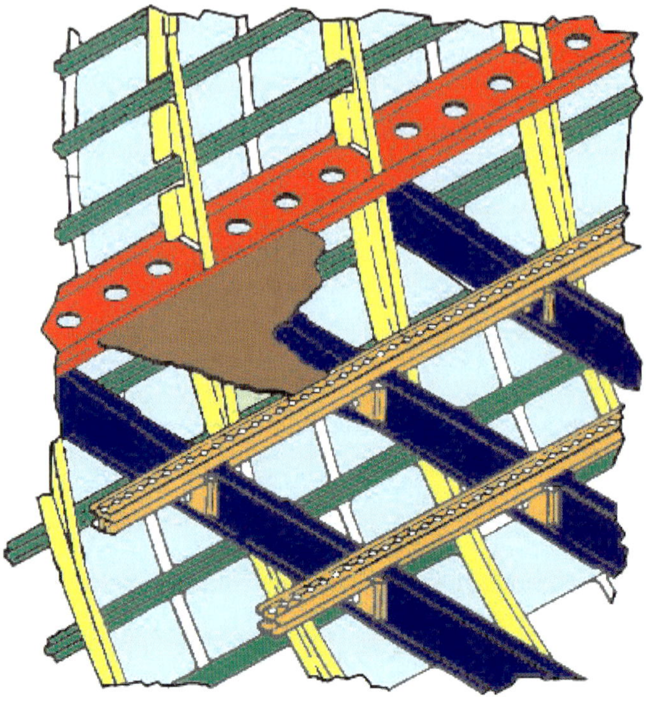

Fig. 5.4

The aircraft's structure is kept as light as possible; the floor is no exception. Lightweight panels are supported on crossbeams (shown in blue in the figure above). These are attached to the longitudinal members (red), called longerons, and the circular frames (yellow). In the passenger cabin, seat rails (dark yellow) carry the passengers seated mass through to the crossbeams. As the structure is light, the loader must take into account the limitations imposed on how much mass may be placed on it.

These are:

- Running load is the total mass allowed over a length of floor.
- Static load is the maximum mass allowed on a given area of a floor.

Examples of how the running load and static load affect an aircraft are given in the CAP 696 data sheets for each aircraft. Below is an example of a Cargo Compartment Limitations table.

The hold has an overall length of 280 inches and is located between balance arms 220 to 500 in. The hold is sub-divided into three separate sections: from 220 to 280 in, from 280 to 340 in, and from 340 to 500 in.

FORWARD CARGO COMPARTMENT

	BA – IN 220	280	340	500
MAXIMUM COMPARTMENT RUNNING LOAD IN KG PER INCH	10.5	8.9	15.12	
MAXIMUM DISTRIBUTION LOAD INTENSITY KG PER SQUARE FOOT	60			
MAXIMUM COMPARTMENT LOAD KG	630	534	2419.2	
MAXIMUM TOTAL LOAD KG	3583.2			
HOLD CENTROID B.A.- IN	280			

Referring to the table above, the forward section BA 220 to 280 in (a length of 60 inches), has a maximum running load allowance of 10.5 kg per inch. This means within this section of the hold, each linear inch along the length of the section can support a mass of 10.5 kg. The table shows that the cargo hold is divided into three compartments, each with a different running load limit.

Maximum Compartment Load

The maximum compartment load is determined by the length of the compartment in units of the running load multiplied by the running load. For example, the fwd compartment of the hold is 60 inches long (280 – 220), and the running load is 10.5 kg/in. 60 in X 10.5 kg per inch = 630 kg.

Maximum Total Load

The maximum total load (3583.2 kg) is the sum of the maximum compartment loads for the three compartments.

Running Load

This running load limitation protects the aircraft frame from excessive loads. This is the total load permitted in any length of the aircraft.

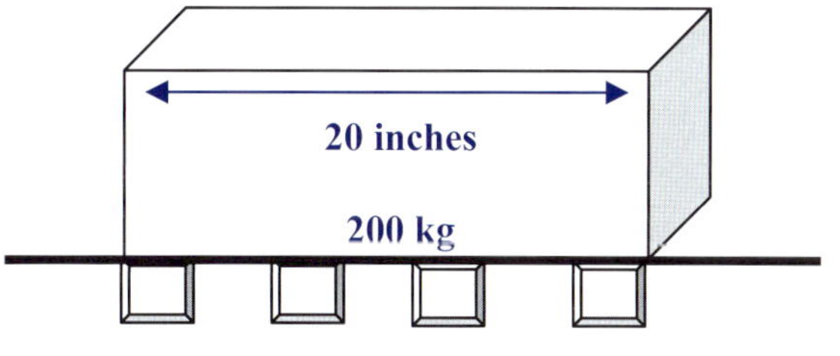

Fig. 5.5

For example, if a container 10 inches wide, 20 inches long, weighing 200 kg is placed in the forward compartment of the hold where the maximum running load is 10.5 kg per inch with its longest length along the hold, the running load is found by dividing the load by the length.

(200 ÷ 20 = 10 kg/in) This is within the limiting load of 10.5 kg.

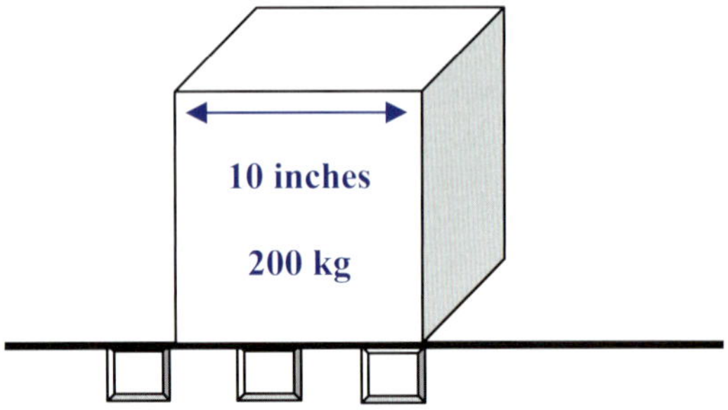

Fig. 5.6

However, if the container is rotated through 90° (as per figure 5.6) the container's running load increases to 200 ÷ 10 = 20 kg/in, which is over the limit.

DISTRIBUTION INTENSITY

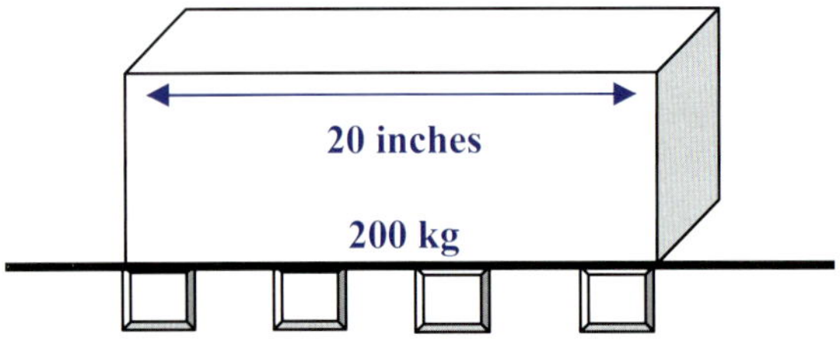

Fig. 5.7

The distribution intensity is the maximum unit mass per square area that the structure can support. For example, a box with a mass of 200 kg and dimensions of 10 in high by 10 in wide by 20 in long exerts an intensity load or square area load of 143.88 kg per square foot.

10 in X 20 in = 200 sq inches
200 sq in ÷ 144 sq inches = 1.39 sq ft [one sq foot = 144 sq inches (12 in X 12 in)]
200 kg ÷ 1.39 sq ft = 143.88 kg per square foot

Therefore, this box could not be placed in the hold as the intensity load far exceeds the 60 kg per square foot limit, whereas the mass of the box (200 kg) is within the running load limit 263 kg (20 X 10.5 = 210 kg). To carry dense small items, load spreaders are used.

Load Spreaders

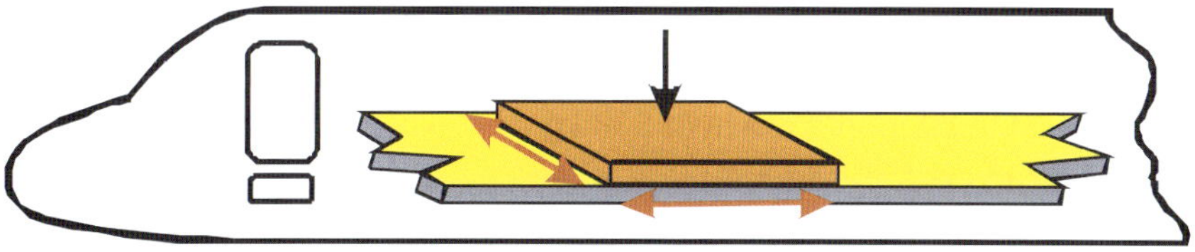

Fig. 5.8

Load spreaders are used to increase the surface area over which a mass can act. For example, if a case (12 in wide by 10 in high by 30 in long) of gold bullion with a mass of 700 kg is to be carried in a hold with a running load of 10 kg per inch and load intensity of 50 kg per square foot, the minimum surface area required to support the load is 14 square feet.

$$700 \text{ kg} \div 50 \text{ kg per sq ft} = 14 \text{ square feet}$$

Therefore, if the bullion is to be carried in this hold, it must have its mass spread over an area of 14 square feet. However, a 14-square-foot area could be a strip 14 ft long by 1 ft wide, or any other combination of lengths and widths that total 14 square feet. To find the minimum length of the load spreader, the running load must be calculated:

$$700 \text{ kg} \div 10 \text{ kg per inch} = 70 \text{ inches, which is } 5.83 \text{ feet}$$

Therefore, to support the load, a load spreader with a minimum area of 14 square feet and side of 70 inches is required. To find the minimum-sized load spreader that could be used, convert the 14 square feet into square inches and divided the answer by 70.

$$14 \text{ sq ft} \times 144 \text{ sq in} = 2016 \text{ sq in}$$
$$2016 \text{ sq in} \div 70 \text{ in} = 28.8 \text{ in}$$

A load spreader of 70 by 28.8 in could be used to support the mass. However, the load spreader itself has a mass that must be taken into account. If the 70 by 28.8 in load spreader has a mass of 40 kg, this must be accounted for within the total load acting over the spreader, as must any equipment that is used to secure the load to the aircraft's floor.

For example:

1. A case of gold bullion (10 in wide by 10 in high by 20 in long) with a mass of 500 kg is to be carried in a hold with a running load of 10 kg per inch and load intensity of 50 kg per square foot. As the intensity of the bullion is greater than the floor's load intensity, which of the following load spreaders is the minimum size that can be used?

   a. 50 in by 30 in with a mass of 30 kg
   b. 40 in by 40 in with a mass of 38 kg
   c. 60 in by 30 in with a mass of 45 kg
   d. 55 in by 35 in with a mass of 48 kg

Solution

Option a. 50 in by 30 in to support a mass of 530 kg
Running load 530 kg ÷ 50 in = 10.6 kg per inch (over limit)

Option b. 40 in by 40 in to support a mass of 538 kg
Running load 538 kg ÷ 40 in = 13.45 kg per inch (over limit)

Option c. 60 in by 30 in to support a mass of 548 kg
Running load 545 kg ÷ 60 in = 9.1 kg per inch (in limit)
Load Intensity = 545 kg ÷ 12.5 sq ft = 43.6 kg per sq ft
60 in x 30 in = 1800 in ÷ by 144 = 12.5 sq ft
(144 is the number of square inches in a sq ft)

Option d. 55 in by 35 in to support a mass of 535 kg
Running load 535 kg ÷ 55 in = 9.73 kg per inch (in limit)
Load Intensity = 535 kg ÷ 11.46 sq ft = 46.7 kg per sq ft

Answer: Option c, as the area is less than that of option d.

CENTROIDS

A centroid is the term used for the balance arm that is said to act for an area. This is normally at the exact centre of the compartment or hold in question. Refer to fig. 5.8 of the hold. The centroid is in the exact centre and is the BA for this hold.

For examples of the static and running load as applied to the MEP 1, turn to page 12 of the CAP 696 MEP 1 data sheet. The bottom line shows the structural floor-loading limit as 120 pounds over an area of one square foot. On page 13 under configuration, each zone is given a maximum load. In zone 3 where 400 pounds can be spread across the whole zone, a package with a base area of 1 square foot and mass of 150 pounds cannot be placed directly onto the aircraft's floor. For small, heavy objects where the static mass would exceed the structural limitation of a floor, load-spreading equipment allows the object's mass to be evenly distributed across a larger area.

SECURITY OF A LOAD

Depending on the aircraft type and the load to be carried, large cargo aircraft use special containers or palletised cargo; passenger aircraft can load baggage into containers or stack luggage directly into the hold. Whichever method is used, the loader must ensure that the load is secure and will not move during flight. Moving loads alters the CG and can cause structural damage or, in the worst case, control problems.

Containers, pallets, load spreaders, and tie down equipment used to secure the load to the aircraft's structure have a mass. These masses have to be taken into account for the mass and balance calculations. These items count as non-revenue load but are part of the traffic load of an aircraft.

ANSWERS

Q1. Calculate the CG and BEM.

As these are single wheel weights, and the aircraft has double wheels on each leg, the given mass for each wheel has to be doubled.

Item	Reaction	Arm	+ Moment kg in	- Moment kg in
Nose Wheels	2000 kg	-92.5 in		-185 000
Left Main Wheels	7500 kg	+68.75 in	+515 625	
Right Main Wheels	7500 kg	+68.75 in	+515 625	
Ballast	-150 kg	+350 in		-52 500
Totals	16 850 kg		+1 031 250	-237 500
			+793 750 kg in	
+793 750 kg in ÷ 16 850 kg = +47.11 in				
BEM = 16 850 kg at a CG of +47 in				

1. From the diagram below, calculate the mass of the aircraft.

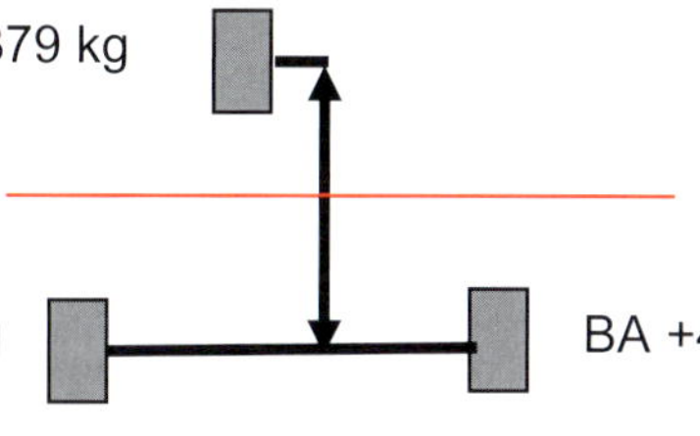

Answer 3386 kg

2. Calculate the CG for the following aircraft

Item	**Reaction**	**Arm**	**+ Moment N in**	**- Moment N in**
Nose Jack	16 677 N	+25 in	+416 925	
Left Main Jack	34 335 N	+303 in	+10 403 505	
Right Main Jack	34 335 N	+303 in	+10 403 505	
Totals	85 347 N		+21 223 935	
+21 223 935 N in ÷ 85 347 N = +248.68 in				
CG of +248.68 in				

Chapter 6
Load Shifting, Load Addition, and Load Subtraction

INTRODUCTION

Chapter 2 showed the basic theory and method of determining the location of the CG. It also discussed how to determine the effect of adding, subtracting, or moving components of the load. As seen in Chapter 2, this becomes time consuming and given that for exam purposes (as well as real life), time is of the essence, a formula is used to calculate both the effect of altering the load or by how much a load must be altered to achieve a desired CG condition.

This formula is shown below, and the rest of this chapter deals with the practical aspects of manipulating this formula. If this formula is not user-friendly, please revert to the default method as per Chapter 2.

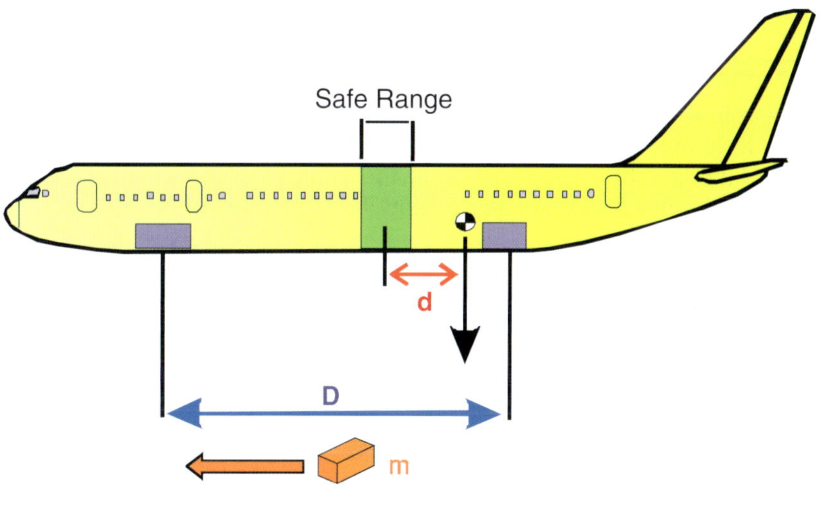

$$^{m}/_{M} = ^{d}/_{D}$$

Where:

- **m** = the mass **to be** moved, added, or subtracted
- **M** = the **total** mass of the aircraft
- **d** = the distance the CG **will** move from its original position
- **D** = the greatest distance that the mass **m** is moved, added to, or subtracted from

LOAD SHIFTING

The act of transferring a mass from one location to another within an aircraft has a double effect on the total moment. The first effect is from removing the mass from one location. Placing this mass in a new location creates the second effect. These complement each other. Calculating the mass to be moved from one location to another always produces a smaller answer than that of removing or adding mass to achieve the same change in CG.

Obviously, the total mass of the aircraft remains constant, as the mass to be shifted is already part of the load. The CG location is likely to alter dependent on the mass moved and the distance it is moved relative to the total mass of the aircraft. The CG always moves toward the position in which the load has been moved to. The formula allows us to find the amount of mass that needs to be shifted, to relocate the CG to a specified point, or the amount by which the CG moves if a known mass is shifted.

Worked examples 1 and 2 show the use of the formula for load shifting.

LOAD ADDITION

Adding any mass to an aircraft has two effects. First, the mass acting over the arm of its location causes a change in the total moments. Second, the gross mass increases. The combined effect results in the CG moving in the direction of the additional load. The result is proportional to the amount of extra load added, the lever arm dimension at which the mass is added, and the total mass of the aircraft.

The formula allows us to find the amount of mass that needs to be added to a given location, to relocate the CG to a specified point, or the amount by which the CG moves if a known mass is added to a given location.

Worked examples 3 and 4 show the use of the formula for load addition.

LOAD SUBTRACTION

Subtracting a mass from an aircraft has the reverse effect to adding extra load, by reducing the total moment and total mass. This results in the CG moving away from the point where the load was removed. This movement is proportional to the amount of the load removed, the lever arm dimension at which the mass had been acting, and the total mass of the aircraft.

The formula is used to find the amount of mass that needs to be removed from a given location to relocate the CG to a given point, or the amount by which the CG moves if a known mass is removed from a given location.

Worked examples 5 and 6 show the use of the formula for load subtraction.

Three practice questions follow each example. Worked answers to each question are at the end of the chapter following the practice questions.

EXAMPLE 1 — FINDING THE MASS TO BE MOVED

Given the following, how much mass must be repositioned to place the CG in the middle of the safe range?

Total mass	10 000 kg
Loaded CG is located at	Stn +8
Fwd limit of the safe range is	Stn +10
Aft limit of the safe range is	Stn +16
Fwd hold is located at	Stn +5
Aft hold is located at	Stn +20

All Stn in feet (ft)

1. Find the location of the intended CG and the distance between it and the existing CG and the direction in which it must move.

2. Find the distance between the location from which the mass is to be removed and the location in which the mass is to be placed.

The line diagram below (which is not to scale) denotes these distances and directions.

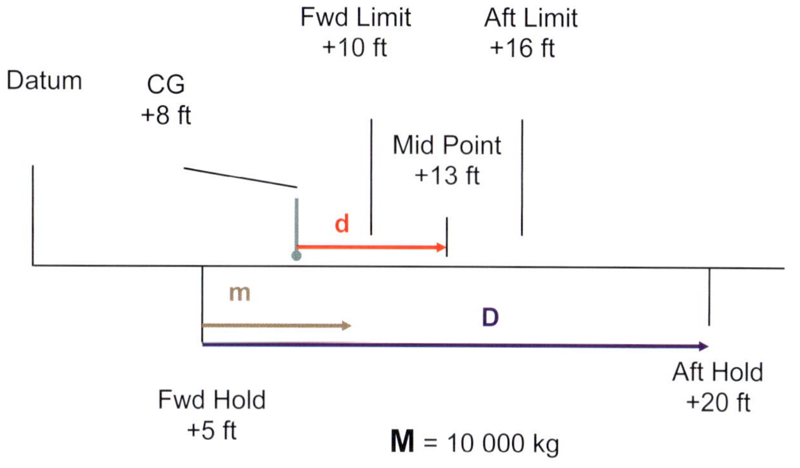

In this example:

m = the mass to **be** moved	**In this case unknown**
M = the **total** mass of the aircraft	10 000 kg
d = the distance the CG **will** move from its original position	5 ft (13 ft – 8 ft)
D = the distance that the mass m **is** moved	15 ft (20 ft – 5 ft)

The new CG location is to be aft of the original location, so m must be moved rearward.

$$\frac{m}{\mathbf{M}} = \frac{d}{\mathbf{D}} \qquad m = \frac{5\text{ ft} \;\; \mathbf{X} \;\; 10\,000\text{ kg}}{15\text{ ft}} \qquad m = \frac{\mathbf{50\,000\text{ ft kg}}}{15\text{ ft}}$$

$$m = \mathbf{3333.333\text{ kg}}$$

Answer: A mass of 3333.33 kg must be repositioned.

PRACTICE 1 — FINDING THE MASS TO BE MOVED

Question 1. How much mass must be moved to put the CG into the middle of the safe range for the aircraft as shown below?

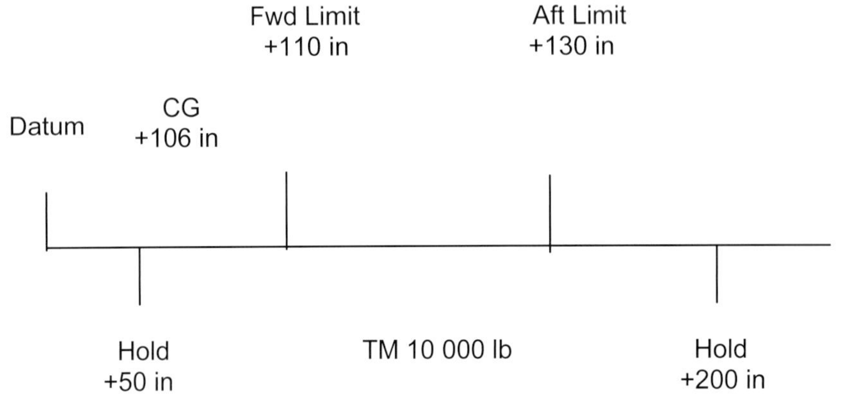

Question 2. How much mass must be transferred to place the following aeroplane in limits?

Datum	Stn	0.0
Fwd limit	Stn	-30.0
Aft Limit	Stn	+25.0
Fwd Hold	Stn	-600.0
Rear Hold	Stn	+600.0
CG located	Stn	+30.0
All Stn in inches		
Total mass		120 000 kg

Question 3. How much mass must be shifted from the front hold to the rear hold of this aircraft to place the CG on the rear limit?

Datum	0.0
Safe Range	6.0 ft
Aft Limit	-2.5 ft
Fwd Hold	-45.0 ft
Rear Hold	+15.0 ft
Total moment	-225 000.00 kg ft
Total mass	50 000 kg

EXAMPLE 2 — FINDING THE EFFECT OF MOVING A KNOWN MASS

The pilot of a light aircraft has loaded the plane with four passengers in the middle and aft row of seats. One of the passengers with a mass of 200 lb seated in the rear row expresses a wish to sit alongside the pilot. Where would the aircraft's CG be located if this is allowed?

Given that:

Total mass as loaded	4451 lb
Loaded CG is located at	+92.0 in
Fwd limit of the safe range is	+82.0 in
Aft limit of the safe range is	+94.0 in
Fwd seats located at	+85.5 in
Mid seats located at	+118.5 in
Aft seats located at	+157.5 in

Method

1. Find the distance between the location from which the known mass is to be removed and the location in which this mass is to be placed.
2. Note the direction in which the known mass is moving.

The line diagram below, which is not to scale, denotes these distances and directions.

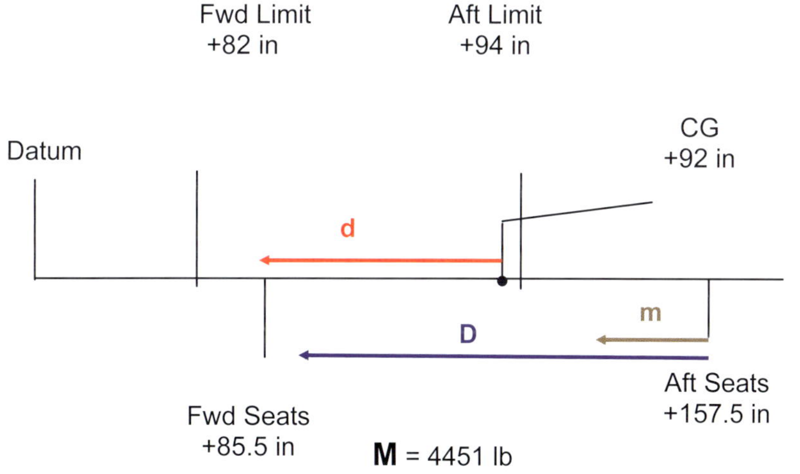

$$d = \frac{m \times D}{M}$$

In this example:

m = the mass **to be** moved	**200 lb**
M = the **total** mass of the aircraft	4451 lb
d = the distance the CG **will** move from its original position	**In this case unknown**
D = the distance that the mass m **is** moved	72 in (157.5 in – 85.5 in)

The original location of mass **m** is aft of the new location for **m**, so **d** must be a forward movement. The value of **d** is the amount by which the CG moves forward.

$$\frac{m}{M} = \frac{d}{D} \qquad d = \frac{200\text{ lb} \times 72\text{ in}}{4451\text{ lb}} \qquad d = \frac{14\,400\text{ lb in}}{4451\text{ lb}}$$

$$d = 3.235\text{ in}$$

The CG moves forward by 3.24 in from its original location to +88.76 in.

PRACTICE 2 — FINDING THE EFFECT ON THE CG WHEN A KNOWN MASS IS MOVED

Question 1. Find the new CG for this aircraft if 100 lb of freight is moved from the rear hold to the forward hold.

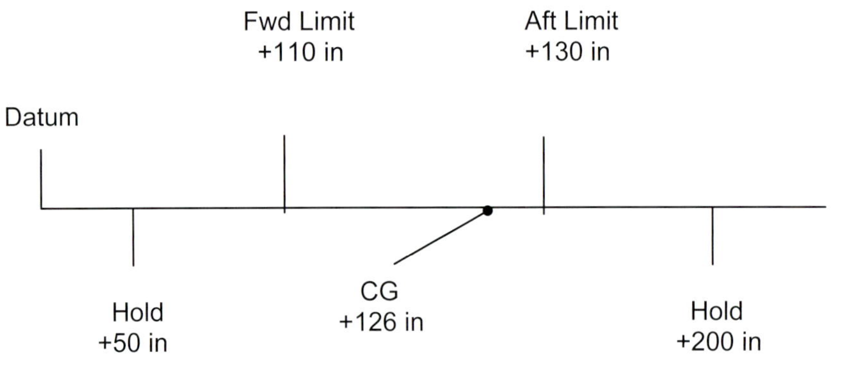

TM 10 000 lb

Question 2. 500 kg is to be relocated into the aft hold from the fwd hold. Will this place this aeroplane in limits?

Datum	Stn	0.0
Fwd limit	Stn	-30.0
Aft Limit	Stn	+25.0
Fwd Hold	Stn	-500.0
Rear Hold	Stn	+200.0
CG located	Stn	-27.0
All Stn in inches (in)		
Total mass		120 000 kg

Question 3. A 1000 kg pallet of freight that was to be loaded into the Fwd hold is found to exceed the cargo door dimensions. It now has to be loaded into the aft hold. Determine the new CG from the original loading data below.

Datum	0.0
Safe Range	6.0 ft
Aft Limit	-2.5 ft
Fwd Hold	-45.0 ft
Rear Hold	+15.0 ft
Total moment	-225 000.00 kg ft
Total mass	50 000 kg

EXAMPLE 3 — FINDING THE MASS THAT MUST BE ADDED TO ALTER THE CG LOCATION

How much mass must be added to the rear hold to place the CG in the middle of the safe range?

Given that:

Total mass of	10 000 kg
Loaded CG is located at	Stn +8
Fwd limit of the safe range is	Stn +10
Aft limit of the safe range is	Stn +16
Fwd Hold is located at	Stn +5
Aft Hold is located at	Stn +20
All Stn in feet (ft)	

Method:

1. Find the location of the intended CG, the distance between it and the existing CG, and the direction in which it must move. This will give the value of **d**.

2. Find the distance between the new CG location and the hold in which the mass is to be added. This will give the value of **D**.

3. As the aircraft's new total mass will not be known until **m** is determined, the original total mass **M** is used.

4. Use the equation **m /M = d/D** as shown on the following page.

The line diagram below, which is not to scale, denotes these distances and directions.

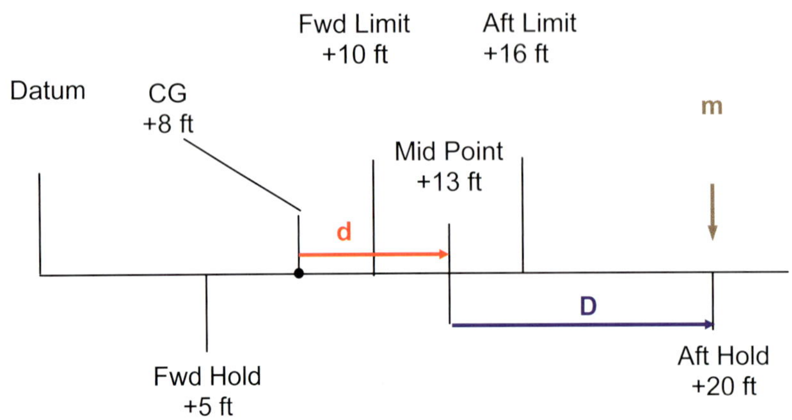

M = 10 000 kg

In this example:

m = the mass **to be** added	**In this case unknown**
M = the original **total** mass of the aircraft	10 000 kg
d = the distance the CG **will** move from its original position	5 ft (13 ft – 8 ft)
D = the distance that the mass **m is** added (from the new CG position)	7 ft (20 ft – 13 ft)

As the new CG location is given, the arm **D** is the greatest distance from this location that **m** can be added. In this case, the aft hold is given.

$$\frac{m}{\mathbf{M}} = \frac{d}{D} \qquad m = \frac{5\text{ ft} \ \mathbf{X} \ 10\,000\text{ kg}}{7\text{ ft}} \qquad m = \frac{\mathbf{50\,000\ ft\ kg}}{7\text{ ft}}$$

$$m = \mathbf{7142.86\ kg}$$

7142.9 kg would have to be added to the aft hold to reposition the CG.

PRACTICE 3 — FINDING THE MASS TO BE ADDED

Question 1. How much mass must be added to put the CG into the middle of the safe range for the aircraft as shown below?

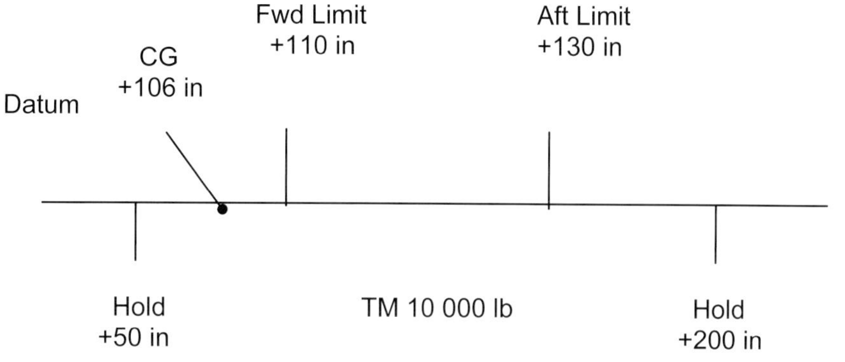

Question 2. How much mass must be added to place the following aeroplane in limits?

Datum	Stn	0.0
Fwd limit	Stn	-30.0
Aft Limit	Stn	+25.0
Fwd Hold	Stn	-600.0
Rear Hold	Stn	+600.0
CG located	Stn	+30.0
All Stn in inches		
Total mass		120 000 kg

Question 3. How much additional mass must placed in a hold of this aircraft to relocate the CG on the forward limit?

Datum	0.0
Safe Range	6.0 ft
Aft Limit	-2.5 ft
Fwd Hold	-45.0 ft
Rear Hold	+15. 0 ft
Total moment	-225 000.00 kg ft
Total mass	50 000 kg

EXAMPLE 4 — FINDING THE EFFECT OF ADDING A KNOWN MASS

The pilot of a light aircraft has worked out the loading of the aircraft for a flight with four passengers in the middle and aft row of seats. Before take-off, another pilot with a mass of 185 lb asks to occupy the front seat. Calculate the aircraft's new CG if the second pilot seat is taken.

Given that:

Total mass as loaded	4451 lb
Loaded CG is located at	+92.0 in
Fwd limit of the safe range is	+82.0 in
Aft limit of the safe range is	+94.0 in
Fwd seats located at	+85.5 in
Mid seats located at	+118.5 in
Aft seats located at	+157.5 in

In this calculation, as the mass to be added 'm' is known, **M** becomes the new total mass.

Method

1. Find **D**, the distance between the location of the mass to be added and the current CG.
2. Find **M**, the original total mass plus **m = M** the new total mass.
3. Note the direction in which the known mass is being added relative to the existing CG.

The line diagram below, which is not to scale, denotes these distances and directions.

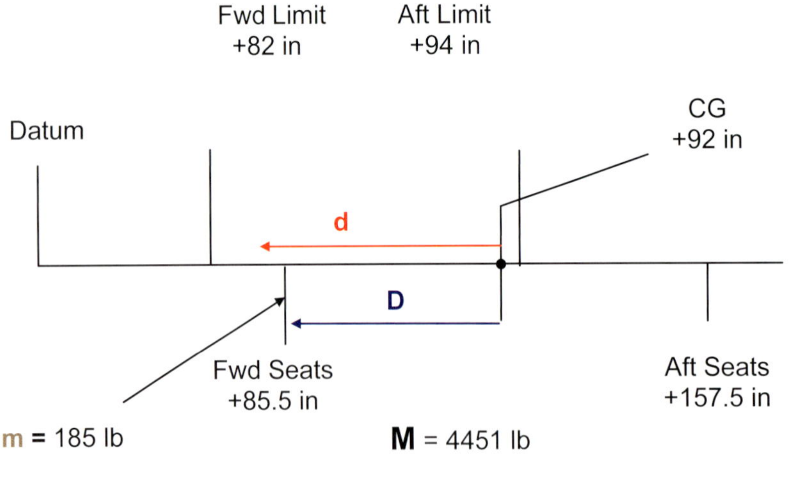

In this example:

m = the mass **to be** added	**185 lb**
M = the new **total** mass of the aircraft	4636 lb (4451 lb + 185 lb)
d = the distance the CG **will** move from its original position	**In this case unknown**
D = the distance that the mass 'm' **is** added from the old CG	6.5 in (92 in – 85.5 in)

As the original CG is known, **D** becomes the distance between it and the point at which **m** is being added. As this is in front of the original CG, **d** is the amount by which the CG moves forward.

$$\frac{m}{M} = \frac{d}{D} \qquad d = \frac{185\ \text{lb} \times 6.5\ \text{in}}{4636\ \text{lb}} \qquad d = \frac{1202.5\ \text{lb in}}{4636\ \text{lb}}$$

$$d = 0.259\ \text{in}$$

Therefore, the new centre of gravity is 92.0 in – 0.26 in = 91.74 in

PRACTICE 4 — FINDING THE NEW CG AFTER A KNOWN MASS IS ADDED

Question 1. What is the effect on this aircraft's CG if 300 lb of baggage is stowed in the cabin at a location of +120.5 in?

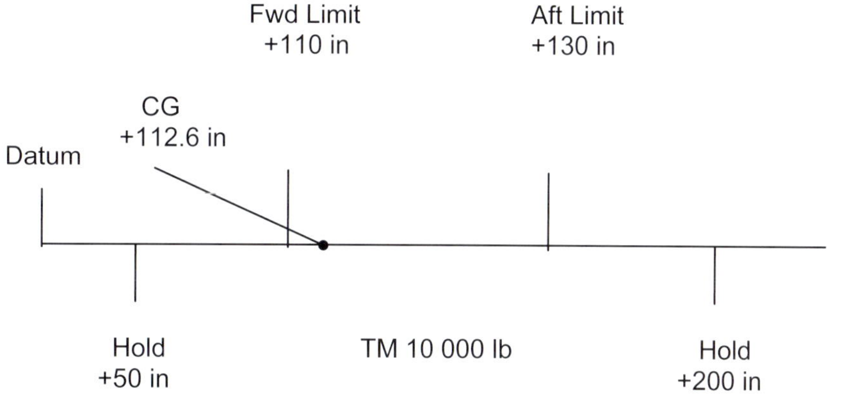

Question 2. If a mass of 3000 kg is to be loaded into the aft hold, is the aircraft in limits?

Datum	Stn	0.0
Fwd limit	Stn	-30.0
Aft Limit	Stn	+25.0
Fwd Hold	Stn	-600.0
Rear Hold	Stn	+600.0
CG located	Stn	-1.0
All Stn in inches		
Total mass		120 000 kg

Question 3. An item of freight with a moment effect of -135 000 kg ft is added to the fwd hold. What is the new CG?

Datum	0.0
Safe Range	6.0 ft
Aft Limit	-2.5 ft
Fwd Hold	-45.0 ft
Rear Hold	+15. 0 ft
Total moment	-225 000.00 kg ft
Total mass	50 000 kg

EXAMPLE 5 — FINDING THE MASS THAT MUST BE REMOVED TO ALTER THE CG LOCATION

How much mass must be subtracted from the fwd hold to place the CG in the middle of the safe range?

Given that:

Total mass of		10 000 kg
Loaded CG is located at	Stn	+8
Fwd limit of the safe range is	Stn	+10
Aft limit of the safe range is	Stn	+16
Fwd Hold is located at	Stn	+5
Aft Hold is located at	Stn	+20
All Stn in feet (ft)		

Method

1. Find the location of the intended CG, the distance between it and the existing CG, and the direction in which the CG must move.

2. Find **D**, the distance between the new CG location and the hold in which the mass is to be removed from.

3. As the aircraft's new total mass is not known until **m** is determined, **M**, the original total mass, is used.

The line diagram below, which is not to scale, denotes these distances and directions:

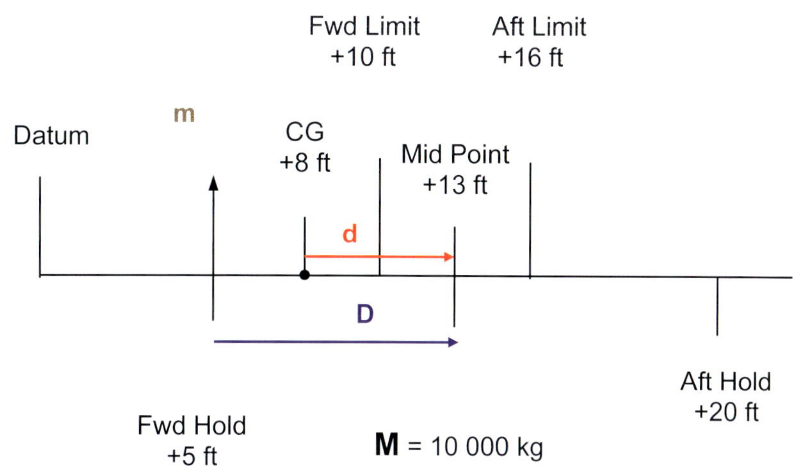

In this example:

m = the mass **to be** removed	**In this case unknown**
M = the original **total** mass of the aircraft	10 000 kg
d = the distance the CG **will** move from its original position	5 ft (13 ft – 8 ft)
D = the distance that the mass m **is** removed from the new CG	8 ft (13 ft – 5 ft)

The result of moving a mass directly on to the CG is for the mass to have no moment effect. The new CG location is given, and the arm D is the distance between the location from where m is removed from and the new CG location.

$$\frac{m}{M} = \frac{d}{D} \qquad m = \frac{5\text{ ft} \times 10\,000\text{ kg}}{8\text{ ft}} \qquad m = \frac{50\,000\text{ kg ft}}{8\text{ ft}}$$

$$m = \mathbf{6250.0\ kg}$$

6250.0 kg would have to be removed from the fwd hold to reposition the CG.

PRACTICE 5 — FINDING THE MASS TO BE REMOVED

Question 1. How much mass must be deducted to locate the CG into the middle of the SR of the aircraft as shown below?

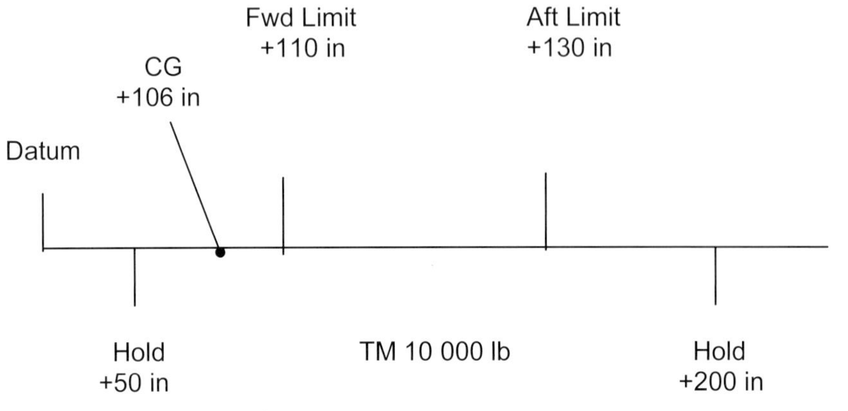

Question 2. How much mass must be removed to place the following aeroplane in limits?

Datum	Stn 0.0
Fwd limit	Stn -30.0
Aft Limit	Stn +25.0
Fwd Hold	Stn -600.0
Rear Hold	Stn +600.0
CG located	Stn +30.0
All Stn in inches (in)	
Total mass	120 000 kg

Question 3. How much mass must be extracted from this aircraft to place its CG on the aft limit?

Datum	0.0
Safe Range	6.0 ft
Aft Limit	-2.5 ft
Fwd Hold	-45.0 ft
Rear Hold	15. 0 ft
Total moment	-225 000.00 kg ft
Total mass	50 000 kg

EXAMPLE 6 — FINDING THE EFFECT OF REMOVING A KNOWN MASS

The pilot of a light aircraft has calculated the loading of the aircraft for a flight with four passengers into the middle and aft row of seats. Before take-off, the two passengers in the rear row cancel their trip. Their combined mass was 294 lb. Calculate the aircraft's CG if the other passengers remain in the middle seats.

Given that:

Total mass as loaded	4451 lb
Loaded CG is located at	+92.0 in
Fwd limit of the safe range is	+82.0 in
Aft limit of the safe range is	+94.0 in
Fwd seats located at	+85.5 in
Mid seats located at	+118.5 in
Aft seats located at	+157.5 in

Method

1. Find **D** the distance between the location of the mass to be removed and the original CG.

2. Find **M,** the original total mass minus **m** **= M** the new total mass.

3. Note the direction in which **m** is being removed from relative to the existing CG.

The line diagram below, which is not to scale, denotes these distances and directions:

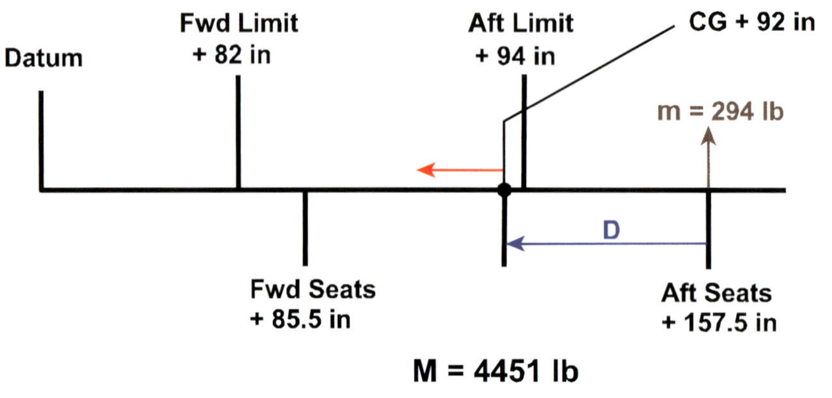

In this example:

m = the mass **to be** removed	294 lb
M = the new **total** mass of the aircraft	4157 lb (4451 lb – 294 lb)
d = the distance the CG **will** move from its original position	**In this case unknown**
D = the furthest distance that the mass m **can** move	65.5 in (157.5 – 92 in)

As the original CG is known, **D** becomes the distance between it and the point from which **m** is being removed. As this is aft of the original CG, **d** is the amount by which the CG moves forward.

$$\frac{m}{M} = \frac{d}{D} \qquad d = \frac{294 \text{ lb} \times 65.5 \text{ in}}{4157 \text{ lb}} \qquad d = \frac{19\,257 \text{ lb in}}{4157 \text{ lb}}$$

$$d = 4.632 \text{ in}$$

The CG will move forward by 4.63 in from its original location to +87.37 in.

PRACTICE 6 — FINDING THE NEW CG AFTER A KNOWN MASS IS REMOVED

Question 1. What is the effect on this aircraft's CG if 250 lb of baggage is off loaded from the aft hold?

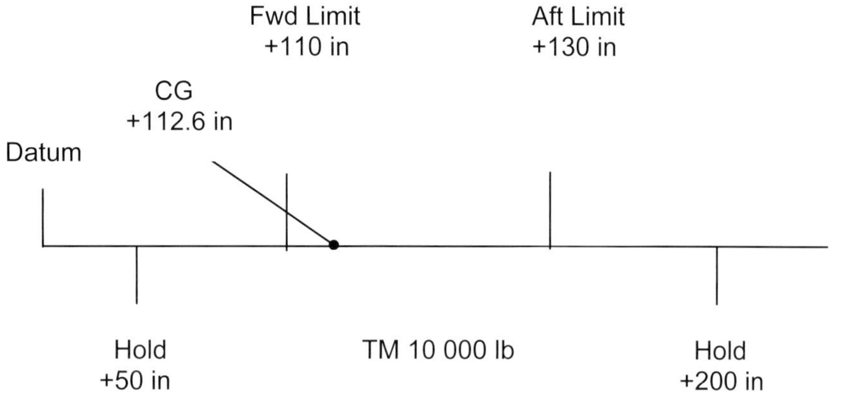

Question 2. If a mass of 1000 kg is to be unloaded from the fwd hold, is the aircraft in limits?

Datum	Stn	0.0
Fwd limit	Stn	-30.0
Aft Limit	Stn	+25.0
Fwd Hold	Stn	-600.0
Rear Hold	Stn	+600.0
CG located	Stn	-1.0
All Stn in inches		
Total mass		120 000 kg

Question 3. An item of freight removed from the aircraft's rear hold has a moment effect of -1670 kg ft, what will be the new CG?

Datum	0.0
Safe Range	6.0 ft
Aft Limit	-2.5 ft
Fwd Hold	-45.0 ft
Rear Hold	+15. 0 ft
Total moment	-225 000.00 kg ft
Total mass	50 000 kg

ANSWERS FOR PRACTICE QUESTIONS

PRACTICE 1 — QUESTION 1

The line diagram below, which is not to scale, denotes these distances and directions:

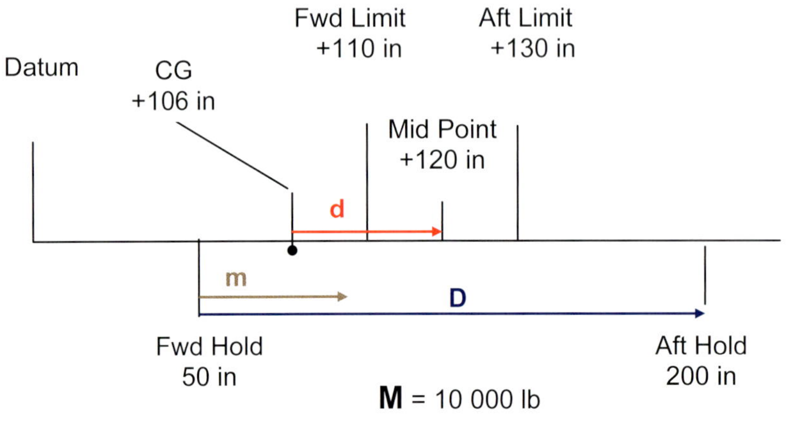

m = the mass **to be** moved	**In this case unknown**
M = the **total** mass of the aircraft	10 000 lb
d = the distance the CG **will** move from its original position	14 in (120 in – 106 in)
D = the distance that the mass m **is** moved	150 in (200 in – 50 in)

The new CG location is to be aft of the original location, so m must be moved rearward.

$$\frac{m}{M} = \frac{d}{D} \qquad m = \frac{14\text{ ft} \times 10\,000\text{ lb}}{150\text{ in}} \qquad m = \frac{140\,000\text{ lb in}}{150\text{ in}}$$

$$m = 933.33\text{ lb}$$

933.33 lb would have to be moved from the Fwd hold to the aft hold.

PRACTICE 1 — QUESTION 2

The line diagram below, which is not to scale, denotes these distances and directions:

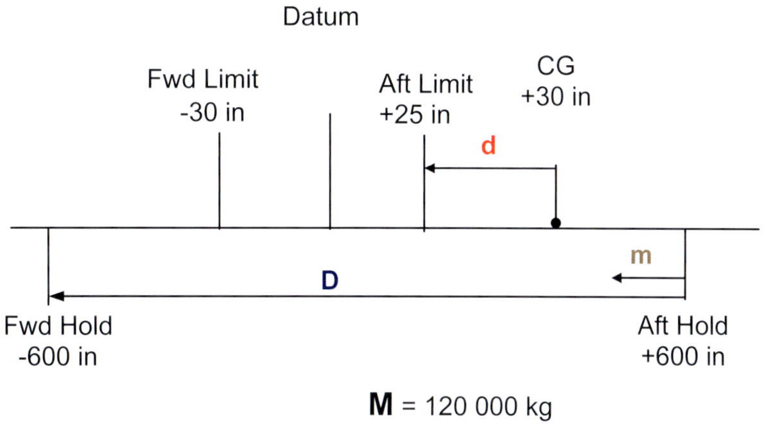

m = the mass **to be** moved	**In this case unknown**
M = the **total** mass of the aircraft	120 000 kg
d = the distance the CG **will** move from its original position	5 in (30 in – 25 in)
D = the distance that the mass m **is** moved	1200 in (600 in + 600 in)

The new CG location is to be forward of the original location, so m must be moved forward.

$$\frac{m}{M} = \frac{d}{D} \qquad m = \frac{5 \text{ in} \times 120\,000 \text{ kg}}{1200 \text{ in}} \qquad m = \frac{600\,000 \text{ kg in}}{1200 \text{ in}}$$

$$m = 500 \text{ kg}$$

500 kg would have to be moved from the aft hold to the Fwd hold.

PRACTICE 1 — QUESTION 3

The line diagram below, which is not to scale, denotes these distances and directions:

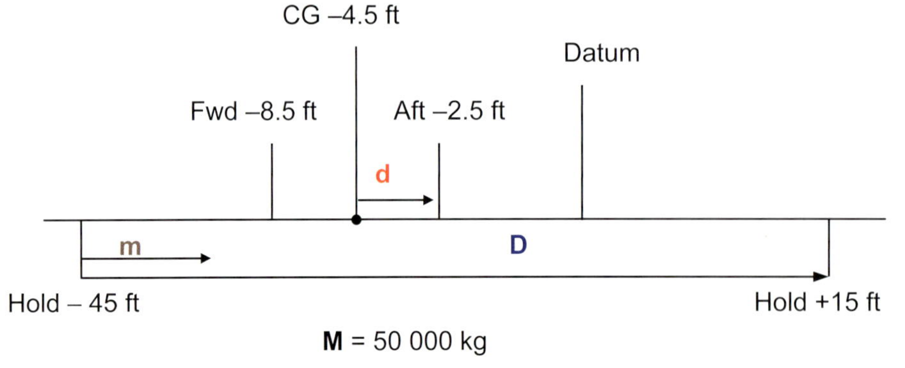

m = the mass **to be** moved	**In this case unknown**
M = the **total** mass of the aircraft	50 000 kg
d = the distance the CG **will** move from its original position	2 ft (4.5 ft-2.5 ft)
D = the distance that the mass m **is** moved	60 ft (45 ft +15 ft)

The new CG location is to be aft of the original location, so m must be moved aft.

$$\frac{m}{\mathbf{M}} = \frac{d}{D} \qquad m = \frac{\mathbf{2\ ft \quad X \quad 50\ 000\ kg}}{\mathbf{60\ ft}} \qquad m = \frac{\mathbf{100\ 000\ kg\ in}}{\mathbf{60\ ft}}$$

$$m = \mathbf{1666.666\ kg}$$

1666.67 kg would have to be moved from the Fwd hold to the aft hold.

PRACTICE 2 — QUESTION 1

The line diagram below, which is not to scale, denotes these distances and directions:

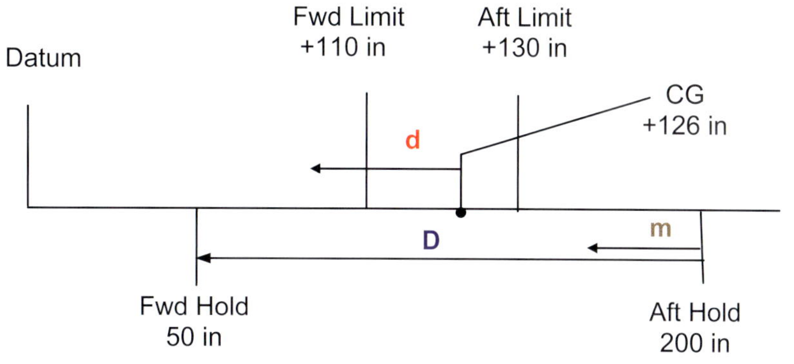

M = 10 000 lb

m = the mass **to be** moved	**100 lb**
M = the **total** mass of the aircraft	10 000 lb
d = the distance the CG **will** move from its original position	**In this case unknown**
D = the distance that the mass **m is** moved	150 in (200 in – 50 in)

As a mass of 100 lb is being moved Fwd, the new CG is Fwd of the current position.

$$\frac{m}{M} = \frac{d}{D} \qquad d = \frac{100\text{ lb} \times 150\text{ in}}{10\,000\text{ lb}} \qquad d = \frac{15\,000\text{ lb in}}{10\,000\text{ lb}}$$

$$d = -1.5\text{ in}$$

CG = +124.5 in (+126 in – 1.5 in)

The new CG would move Fwd by 1.5 in to a location of +124.5 in.

PRACTICE 2 — QUESTION 2

The line diagram below, which is not to scale, denotes these distances and directions.

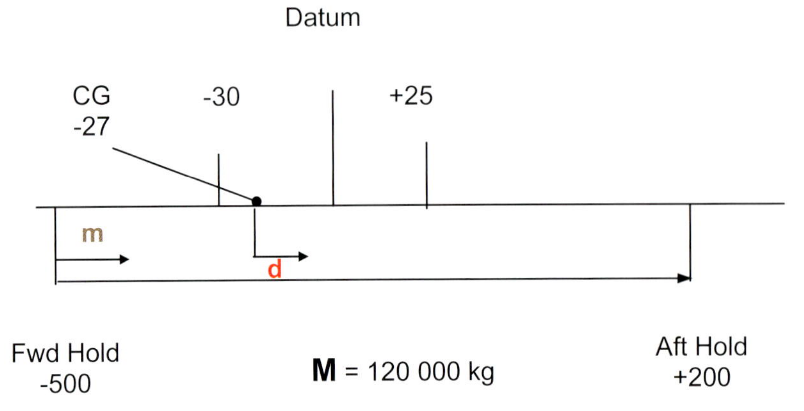

m = the mass **to be** moved — **500 kg**
M = the **total** mass of the aircraft — 120 000 kg
d = the distance the CG **will** move from its original position — **In this case unknown**
D = the distance that the mass m **is** moved — 700 in (500 in +200 in)

A mass of 500 kg is to be moved into the aft hold from the Fwd hold, so the CG moves aft.

$$\frac{m}{M} = \frac{d}{D} \qquad d = \frac{500\text{ kg} \times 700\text{ in}}{120\,000\text{ kg}} \qquad d = \frac{350\,000\text{ kg in}}{120\,000\text{ kg}}$$

$$d = +2.917\text{ in}$$

CG = -24.08 in (-27 + 2.92)

The New CG is located 2.916 inches aft of the original CG position at 24.08 in.

PRACTICE 2 — QUESTION 3

The line diagram below, which is not to scale, denotes these distances and directions.

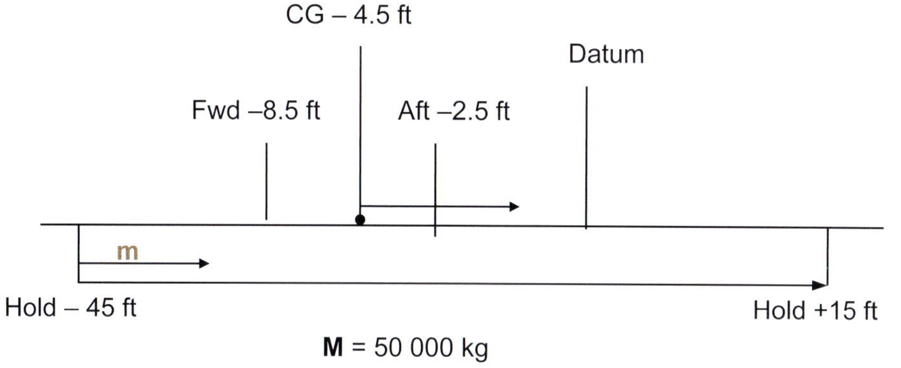

m = the mass **to be** moved	1000 kg
M = the **total** mass of the aircraft	50 000 kg
d = the distance the CG **will** move from its original position	**In this case unknown**
D = the distance that the mass m **is** moved	60 ft (45 ft +15 ft)

As a 1000 kg is to be relocated from the fwd hold to the aft hold, the CG moves aft.

$$\frac{m}{M} = \frac{d}{D} \qquad d = \frac{1000\text{ kg} \times 60\text{ ft}}{50\,000\text{ kg}} \qquad d = \frac{60\,000\text{ kg ft}}{50\,000\text{ kg}}$$

$$d = +1.2\text{ ft}$$

CG = -3.3 ft (-4.5 ft + 1.2 ft)
The new CG is located 1.2 ft aft of the original CG at –3.3 ft which is in limits.

PRACTICE 3 — QUESTION 1

The line diagram below, which is not to scale, denotes these distances and directions.

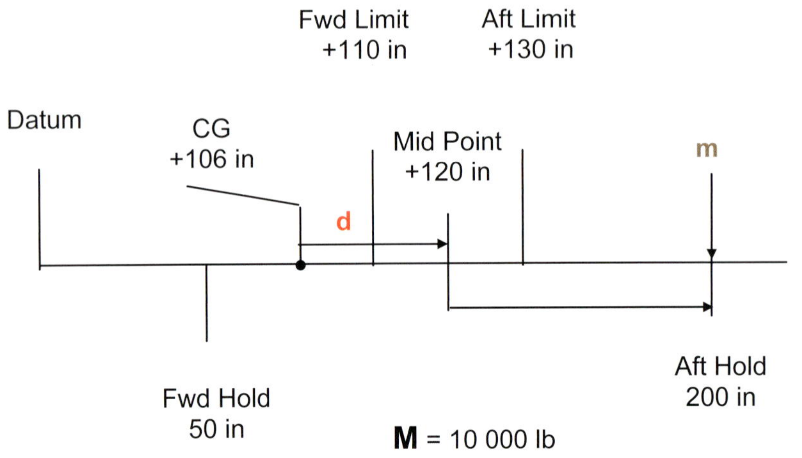

m = the mass **to be** added	**In this case unknown**
M = the original **total** mass of the aircraft	10 000 lb
d = the distance the CG **will** move from its original position	14 in (120 in-106 in)
D = the distance that the mass m **is** moved	80 in (200 in – 120 in)

The new CG location is to be aft of the original location, so m**,** an unknown, must be added to the rear hold, so **D** is between the new CG and the place m is added to.

$$\frac{m}{M} = \frac{d}{D} \qquad m = \frac{14 \text{ in} \times 10\,000 \text{ lb}}{80 \text{ in}} \qquad m = \frac{140\,000 \text{ lb in}}{80 \text{ in}}$$

$$m = 1750 \text{ lb}$$

1750 lb would have to be added to the aft hold.

PRACTICE 3 — QUESTION 2

The line diagram below, which is not to scale, denotes these distances and directions.

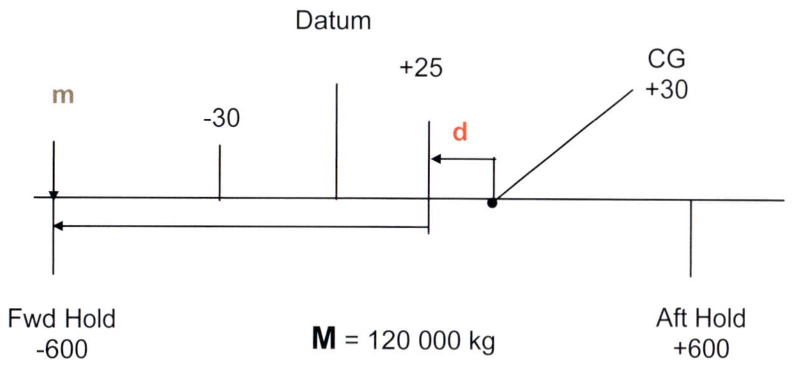

m = the mass **to be** added	**In this case unknown**
M = the original **total** mass of the aircraft	120 000 kg
d = the distance the CG **will** move from its original position	5 in (30 in – 25 in)
D = the distance that the mass m **is** moved	625 in (600 in +25 in)

A mass m kg is to be added into the Fwd hold to move the CG on to the aft Limit. **D** is the distance between them.

$$\frac{m}{M} = \frac{d}{D} \qquad m = \frac{5\text{ in} \times 120\,000\text{ kg}}{625\text{ in}} \qquad m = \frac{600\,000\text{ kg in}}{625\text{ in}}$$

$$m = 960\text{ kg}$$

An addition of 960 kg to the Fwd hold places the CG on the aft limit.

PRACTICE 3 — QUESTION 3

The line diagram below, which is not to scale, denotes these distances and directions.

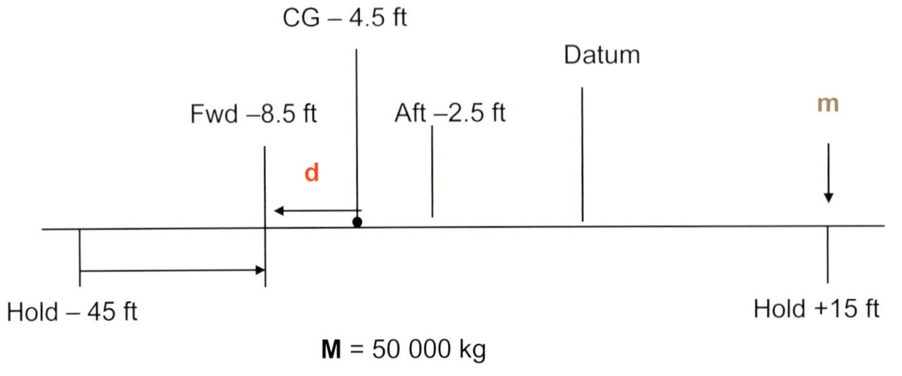

m = the mass **to be** added	**In this case unknown**
M = the original **total** mass of the aircraft	50 000 kg
d = the distance the CG **will** move from its original position	4 ft (8.5 ft –4.5 ft)
D = the distance that the mass 'm' **is** moved, in this case added	36.5 ft (45 ft – 8.5 ft)

As the CG is to be relocated, fwd m must be added ahead of the original CG into the front hold. Therefore, **D** is the distance between the new CG and the hold.

$$\frac{m}{M} = \frac{d}{D} \qquad m = \frac{4\ \text{ft} \times 50\,000\ \text{kg}}{36.5\ \text{ft}} \qquad m = \frac{200\,000\ \text{kg in}}{36.5\ \text{ft}}$$

$$m = 5479.452\ \text{kg}$$

A mass of 5479.45 kg must be added to the Fwd hold to relocate the CG on the Fwd limit of the safe range.

PRACTICE 4 — QUESTION 1

The line diagram below, which is not to scale, denotes these distances and directions.

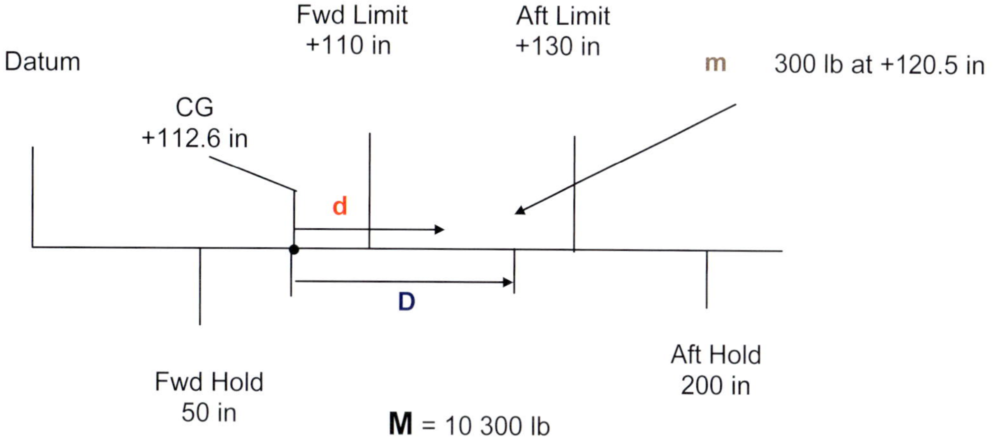

m = the mass **to be** added	**300 lb**
M = the new **total** mass of the aircraft	10 300 lb (10 000 lb +300 lb)
d = the distance the CG **will** move from its original position	**In this case unknown**
D = the distance that the mass **m is** moved	7.9 in (120.5 in – 112.6 in)

The new CG location is aft of the original location as **m,** a 300 lb mass, added to a location behind it, so **D** is between the old CG and the place to which **m** is added.

$$\frac{m}{M} = \frac{d}{D} \qquad d = \frac{300\ \text{lb} \times 7.9\ \text{in}}{10\ 300\ \text{lb}} \qquad d = \frac{2370\ \text{lb in}}{10\ 300\ \text{lb}}$$

$$d = +0.23\ \text{in}$$

CG = +112.83 in (112.6 in + 0.23 in)

The addition of 300 lb at +120.5 in results in the CG relocating to a position of +112.83 in.

PRACTICE 4 — QUESTION 2

The line diagram below, which is not to scale, denotes these distances and directions.

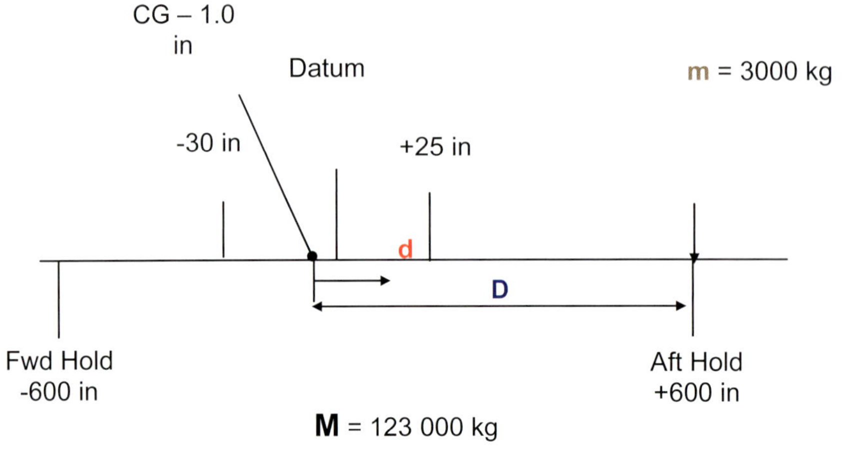

m = the mass **to be** added	**3000 kg**
M = the new **total** mass of the aircraft	123 000 kg (120 000 kg + 3000 kg)
d = the distance the CG **will** move from its original position	**In this case unknown**
D = the distance that the mass m **is** moved	601 in (600 in + 1 in)

A mass m kg is to be added into the aft hold. The CG moves rearward. **D** is the distance between the original CG and aft hold.

$$\frac{m}{M} = \frac{d}{D} \qquad d = \frac{3000\text{ kg} \times 601\text{ in}}{123\,000\text{ kg}} \qquad d = \frac{1\,803\,000\text{ kg in}}{123\,000\text{ kg}}$$

$$d = +14.66\text{ in}$$

CG = +13.66 in (-1.0 in + 14.66 in)
The addition of 3000 kg to the aft hold places the CG at +13.66 in.

PRACTICE 4 — QUESTION 3

The line diagram below, which is not to scale, denotes these distances and directions.

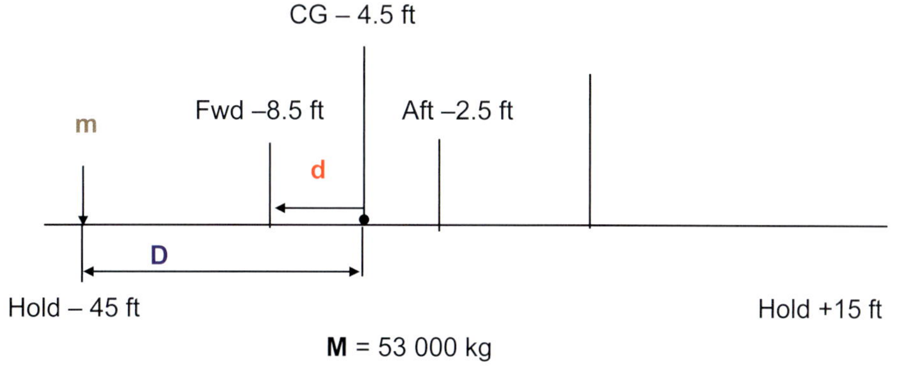

m = the mass **to be** added	**3000 kg**
M = the new **total** mass of the aircraft	53 000 kg (50 000 kg + 3000 kg)
d = the distance the CG **will** move from its original position	**In this case unknown**
D = the distance that the mass m **is** moved	40.5 ft (45 ft – 4.5 ft)

As the mass to be added is given as a moment negative effect, the mass is this effect divided by the arm it is acting over. In this case, -135 000 kg ft ÷ -45 ft = 3000 kg. This mass causes the CG to relocate fwd of the original CG position.

$$\frac{m}{M} = \frac{d}{D} \qquad d = \frac{3000\text{ kg} \times 40.5\text{ ft}}{53\,000\text{ kg}} \qquad d = \frac{121\,500\text{ kg ft}}{53\,000\text{ kg}}$$

$$d = -2.292\text{ ft}$$

CG = -6.79 ft (-4.5 ft + -2.29 ft)

A mass of 3000 kg added to the Fwd hold relocates the CG to a position of -6.79 ft.

PRACTICE 5 — QUESTION 1

The line diagram below, which is not to scale, denotes these distances and directions.

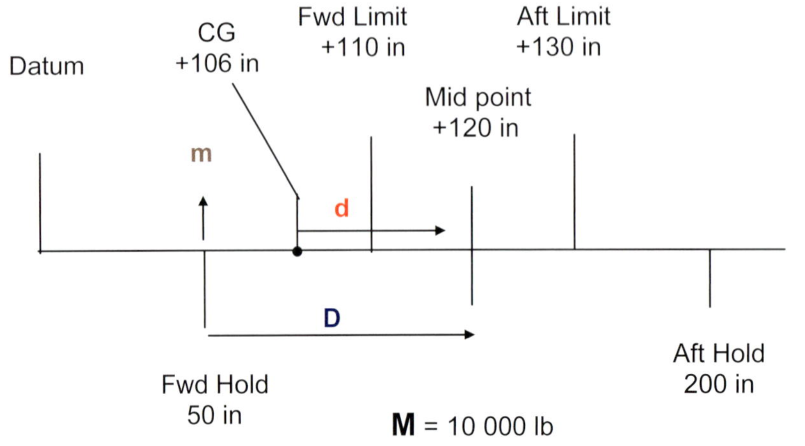

m = the mass **to be** subtracted	**In this case unknown**
M = the original **total** mass of the aircraft	10 000 lb
d = the distance the CG **will** move from its original position	14 in (120 in – 106 in)
D = the distance that the mass **m is** moved	70 in (120 in – 50 in)

The new CG location is aft of the original location as **m,** an unknown mass, is removed from the Fwd hold to relocate the CG by **d** on to the mid point, so **D** is between the hold and new CG.

$$\frac{m}{M} = \frac{d}{D} \qquad m = \frac{14 \text{ in} \quad X \quad 10\,000 \text{ lb}}{70 \text{ in}} \qquad m = \frac{140\,000 \text{ lb in}}{70 \text{ in}}$$

$$m = 2000 \text{ lb}$$

To relocate the CG to the mid point of the safe range, a mass of 2000 lb would have to be removed from the Fwd hold.

PRACTICE 5 — QUESTION 2

The line diagram below, which is not to scale, denotes these distances and directions.

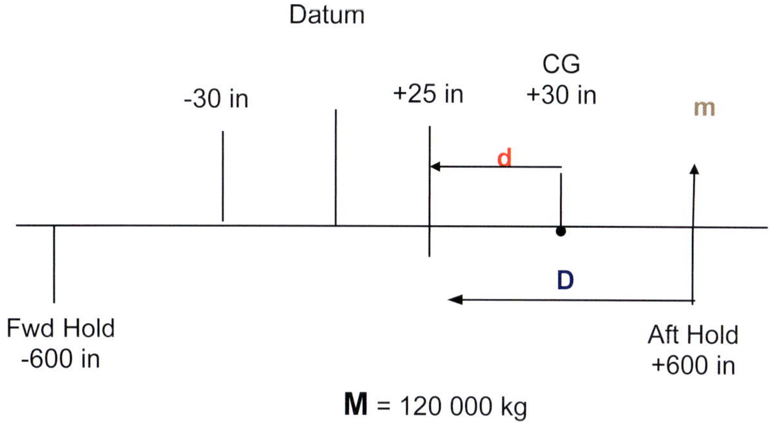

m = the mass **to be** subtracted	**In this case unknown**
M = the original **total** mass of the aircraft	120 000 kg
d = the distance the CG **will** move from its original position	5 in (30 in – 25 in)
D = the distance that the mass m **is** moved	575 in (600 in – 25 in)

A mass m kg is to be removed from the aft hold the CG will move forward, **D** will be the distance between the new CG and aft hold.

$$\frac{m}{M} = \frac{d}{D} \qquad m = \frac{5\text{ in} \times 120\,000\text{ kg}}{575\text{ in}} \qquad m = \frac{600\,000\text{ kg in}}{575\text{ in}}$$

$$m = \mathbf{1043.478\ kg}$$

To relocate the CG to the rear limit, a mass of 1043.48 kg must be removed from the aft hold.

PRACTICE 5 — QUESTION 3

The line diagram below, which is not to scale, denotes these distances and directions.

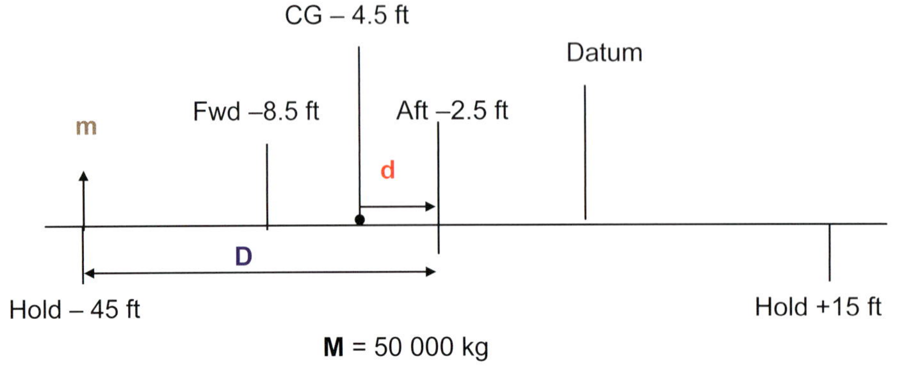

In this example:

m = the mass **to be** subtracted	**In this case unknown**
M = the original **total** mass of the aircraft	50 000 kg
d = the distance the CG **will** move from its original position	2 ft (4.5 ft – 2.5 ft)
D = the distance that the mass m **is** moved	42.5 ft (45 ft – 2.5 ft)

As the mass to be removed is not given but the new location is given, the **D** in this case is between the hold and new CG position.

$$\frac{m}{M} = \frac{d}{D} \qquad m = \frac{2\text{ ft} \times 50\,000\text{ kg}}{42.5\text{ ft}} \qquad m = \frac{100\,000\text{ kg ft}}{42.5\text{ ft}}$$

$$m = 2352.94\text{ kg}$$

A mass of 2352.94 kg must be removed from the Fwd hold to relocate the CG on the aft limit.

PRACTICE 6 — QUESTION 1

The line diagram below, which is not to scale, denotes these distances and directions.

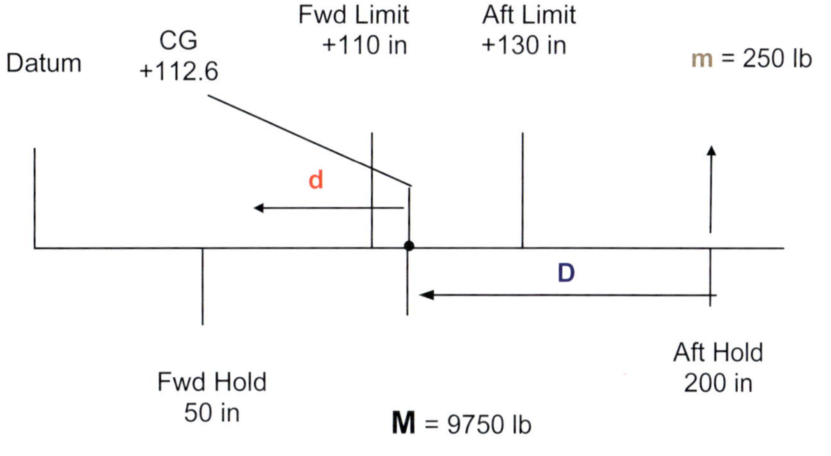

m = the mass **to be** subtracted	**250 lb**
M = the new **total** mass of the aircraft	9750 lb (10 000 lb – 250 lb)
d = the distance the CG **will** move from its original position	**In this case unknown**
D = the distance that the mass **m** **is** moved	87.4 in (200 in – 112.6 in)

As a mass of 250 lb is removed from the aft hold, the new CG is Fwd of the current position.

$$\frac{m}{M} = \frac{d}{D} \qquad d = \frac{250 \text{ lb} \times 87.4 \text{ in}}{9750 \text{ lb}} \qquad d = \frac{21\,850 \text{ lb in}}{9750 \text{ lb}}$$

$$d = -2.241 \text{ in}$$

CG = +110.36 in (+112.6 in – 2.24 in)

The new CG would move Fwd by 2.24 in to a location of +110.36 in.

PRACTICE 6 — QUESTION 2

The line diagram below, which is not to scale, denotes these distances and directions.

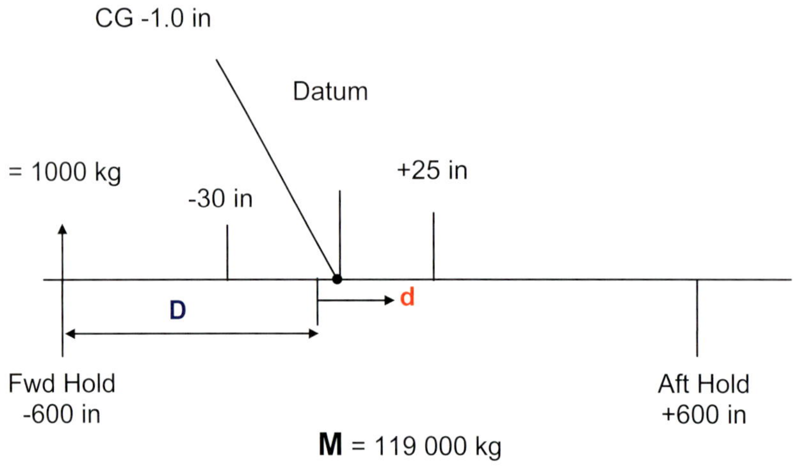

m = the mass **to be** subtracted — **1000 kg**
M = the new **total** mass of the aircraft — 119 000 kg (120 000 kg -1000 kg)
d = the distance the CG **will** move from its original position — **In this case unknown**
D = the distance that the mass **m** **is** removed from the old CG position — 599 in (600 in – 1 in)

As a mass **m** kg is removed from the Fwd hold, the CG moves rearward. **D** is the distance between the original CG and Fwd hold.

$$\frac{m}{M} = \frac{d}{D} \qquad d = \frac{1000\ \text{kg} \times 599\ \text{in}}{119\ 000\ \text{kg}} \qquad d = \frac{599\ 000\ \text{kg in}}{119\ 000\ \text{kg}}$$

$$d = +5.034\ \text{in}$$

CG = +4.03 in (-1.0 in + 5.03 in)
The subtraction of 1000 kg to the Fwd hold places the CG at +4.03 in.

PRACTICE 6 — QUESTION 3

The line diagram below, which is not to scale, denotes these distances and directions.

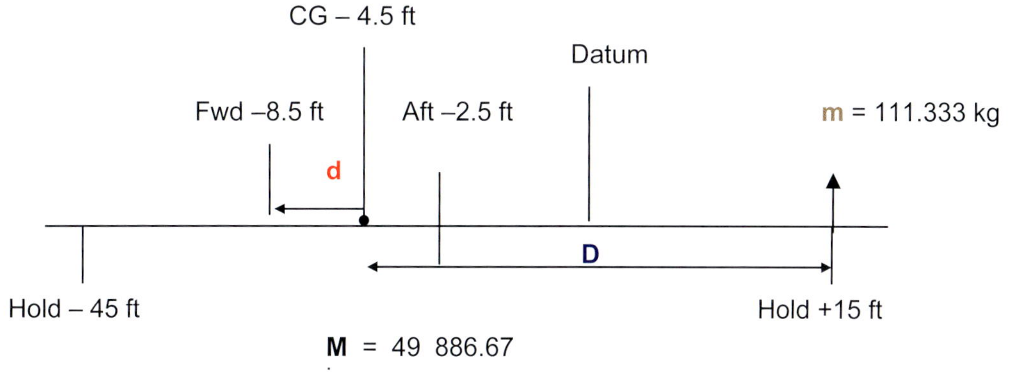

m = the mass **to be** subtracted — **111.333 kg**
M = the new **total** mass of the aircraft — 49 886.67 kg (50 000 kg – 111.333 kg)
d = the distance the CG **will** move from its original position — **In this case unknown**
D = the distance that the mass m **is** moved — 19.5 ft (15 ft + 4.5 ft)

As the mass to be removed has a negative moment effect, the mass is the effect divided by the arm it is acting over. In this case, -1670 kg ft ÷ -15 ft = 111.333 kg. This mass causes the CG to relocate fwd of the original CG position.

$$\frac{m}{M} = \frac{d}{D} \qquad d = \frac{\mathbf{111.333\ kg\ \ X\ \ 19.5\ ft}}{\mathbf{49\ 886.67\ kg}} \qquad d = \frac{\mathbf{2170.99\ kg\ ft}}{\mathbf{49\ 886.67\ kg}}$$

$$d = \mathbf{-0.044\ ft}$$

CG = -4.54 ft (-4.5 ft + -0.04 ft)

A mass of 111.333 kg removed from the aft hold relocates the CG to a position of -4.54 ft.

Chapter 7
Mean Aerodynamic Chord

INTRODUCTION

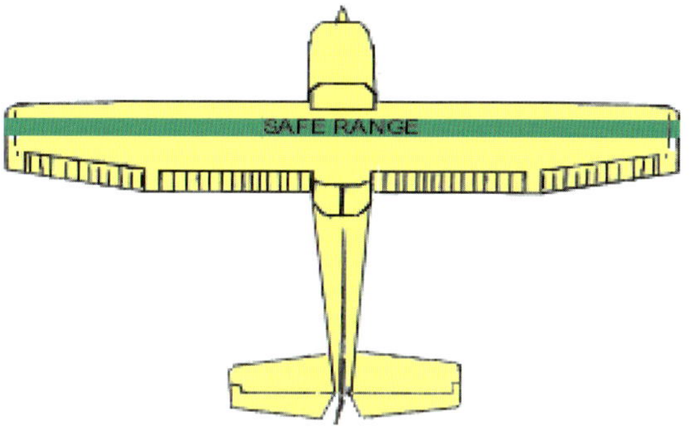

Light aircraft predominantly use straight wings that are either tapered or semi-tapered as per the diagram above. These wings are mounted at right angles to the fuselage. This means that the CG safe range runs across the span of the wings.

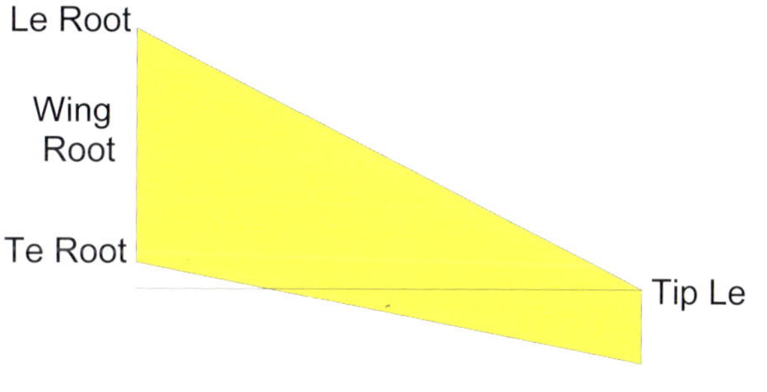

Modern transport aircraft, due to their higher speeds and efficiencies, use swept-back wings. Here, the leading edge (Le) of the wing at the tip is behind the junction of the leading edge at the wing root and in some designs (as shown above) behind the wing root trailing edge (Te). Swept-back wings produce varying degrees of lift across the span of the wings depending on the aircraft's speed and angle of attack.

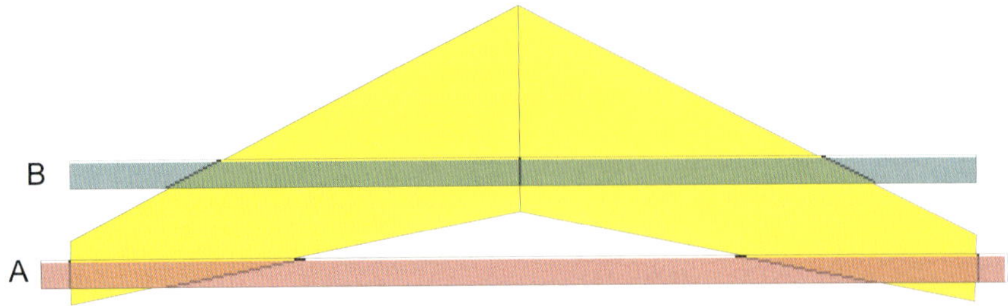

The safe range on a straight wing can easily be visualised as a band running from one wing tip to the other and be seen to fall within the lifting area. For a swept-back wing, if the safe range band were to pass through the wing tips (as per band A in the diagram above), the CG would be behind the Centre of Pressure. To ensure that the CG safe range is ahead of the Centre of Pressure, the safe range is located forward on the wing and is shown above as band B. As can be seen, the safe range effectively exits through the leading edge. To rationalise this, a system called the Mean Aerodynamic Chord (MAC) is used.

MEAN AERODYNAMIC CHORD (MAC)

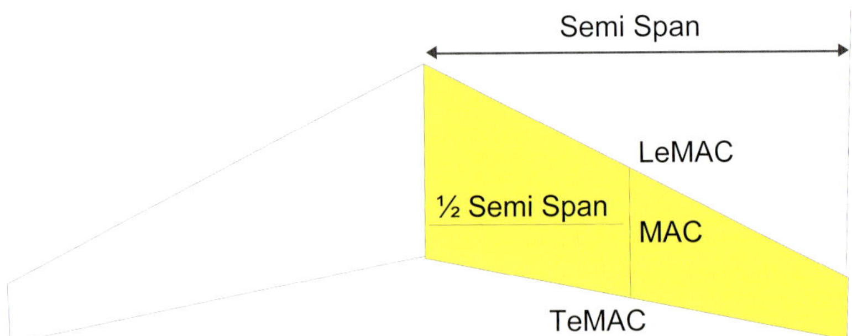

The mean aerodynamic chord of a wing is found by locating the wing's mid semi-span and measuring the chord of the wing at this point. This chord is then considered the mean aerodynamic chord of the aircraft.

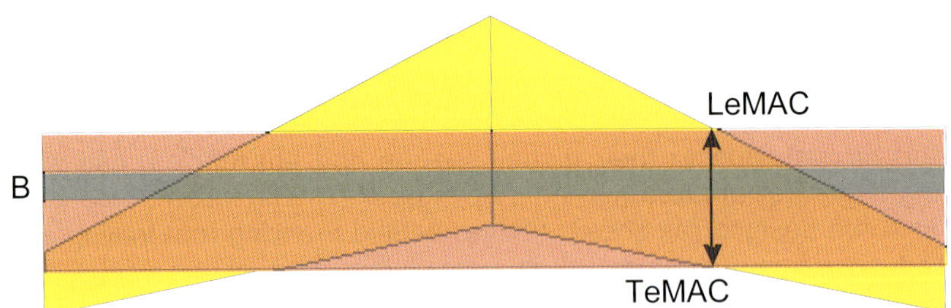

Effective Wing for MAC

For mass and balance purposes, the aircraft can be considered to have a straight wing of this chord and throughout the entire flight envelope. It is considered that the lift generated by the wings comes from this chord. The length of the chord is referred to as MAC. The leading edge of the chord is termed LeMAC and the trailing edge as TeMAC.

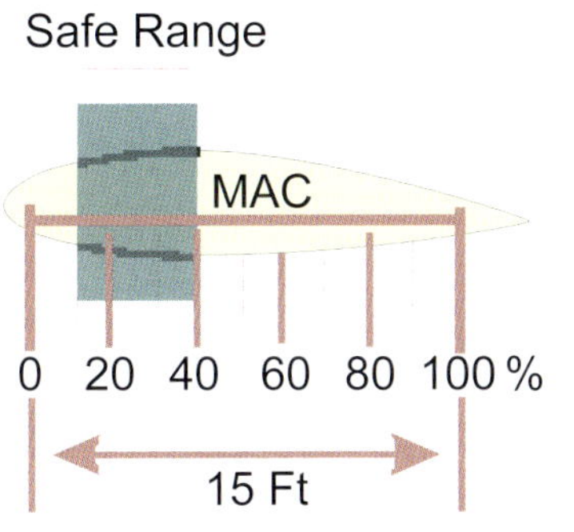

For mathematical computation, whatever the actual length of the chord, it is discussed in the terms of percentage. LeMAC is always 0% MAC, and TeMAC is 100% MAC whatever the distance. This allows the fwd and aft CG limits to be given as percentages of MAC. To find the linear distance that 1% of Mac represents, divide the chord length of MAC by 100.

For example: 15 ft ÷ 100 = 0.15 ft

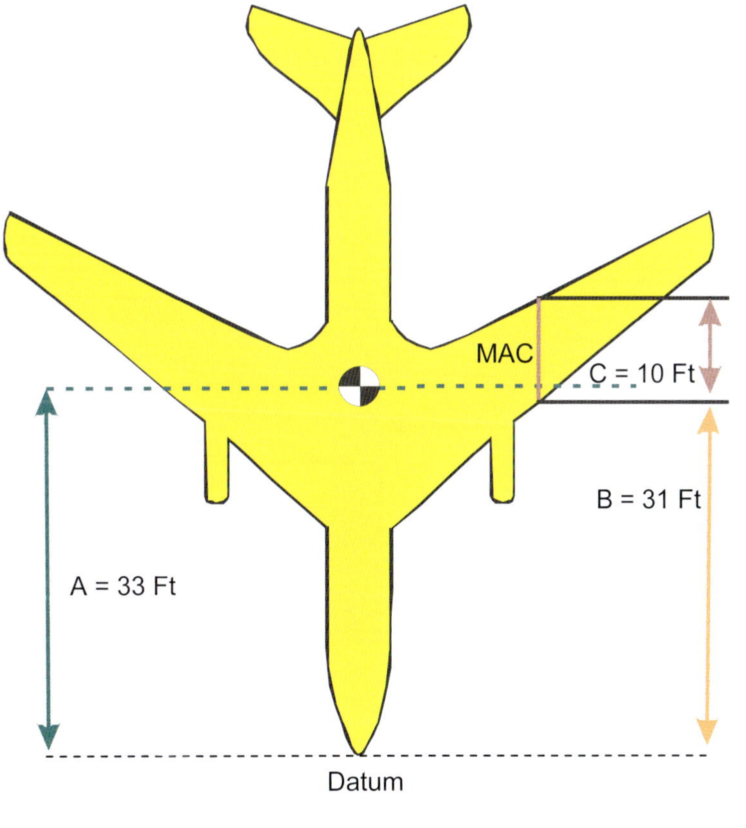

As all components are given in relation to the aircraft's datum, the leading edge of MAC has to be given as a linear distance from the datum. The CG can, both in reality and for exam purposes, be given as a linear distance from the datum that requires converting to a percentage of MAC. Conversely, it can be given as a percentage of MAC that requires converting into a linear distance from the datum.

CONVERTING A LINEAR CG POSITION INTO PERCENTAGE OF MAC

Where the limits and CG locations are given as linear distances from the datum and have to be expressed as % MAC, the following formula can be used:

$$\frac{A - B}{C} \times 100 = \%\ MAC$$

Where:

Where:	A	is the distance from the datum to the CG
	B	is the distance from the datum to LeMAC
	C	is the length of MAC

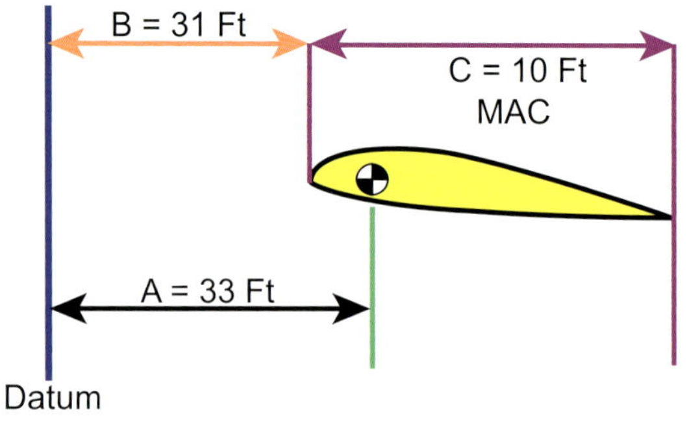

In the example above, the CG is located at 33 ft aft of the datum, the LeMAC is located 31 ft aft of the datum, and MAC is 10 ft.

$$\frac{A - B}{C} \quad X \quad 100 \quad = \quad \% \quad MAC$$

$$\frac{33 - 31}{10} \quad X \quad 100 \quad = \quad 20\% \quad MAC$$

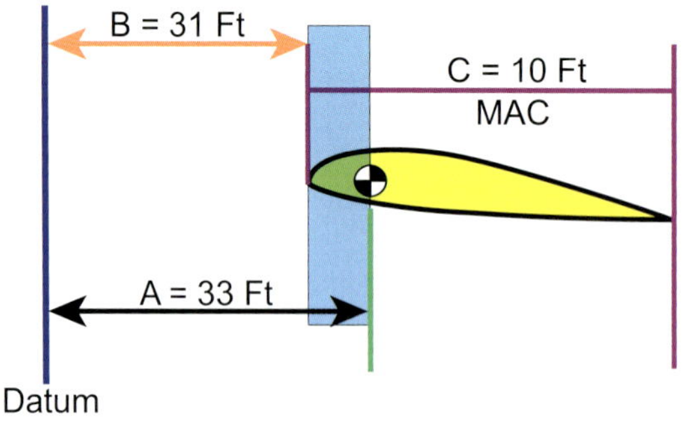

In reality, A minus B is finding the linear distance of the CG from LeMAC 33 ft – 31 ft = 2 ft, shown in the diagram above as a blue band.

$$\frac{A - B}{C} \quad X \quad 100 \quad = \quad \% \quad MAC$$

While the formula is very useful and must be remembered, as it can in itself be the answer to an exam question, it can become very unwieldy when attempting to work out questions where the datum is located within the MAC, as shown in the following example.

The LeMAC is located at -2.5 m, MAC is 3.7 m, and the CG is located at -1.9 m. Find the CG as a percentage of MAC.

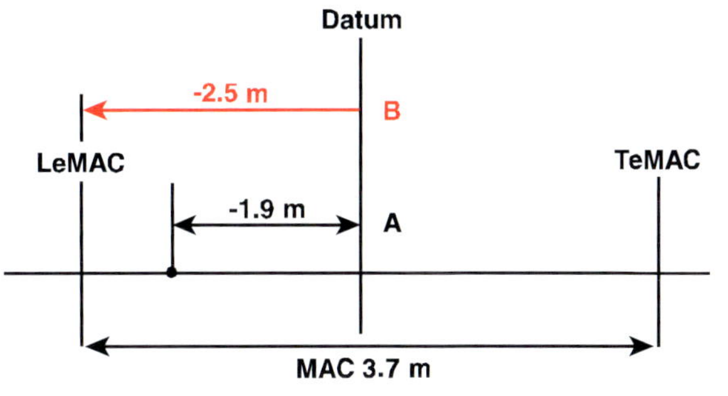

A diagram of the problem, which is not to scale, (as shown above) can help to visualise the problem.

$$\frac{A - B}{C} \times 100 = \% \quad MAC$$

$$\frac{-1.9 - -2.5}{3.7} \times 100 = 16.22\% \quad MAC$$

Here it can be seen that the formula solves the problem, provided that the signs are used as shown above. However, if the signs are omitted by mistake, the answer, while being the same numerical value, becomes negative as shown below.

$$\frac{1.9 - 2.5}{3.7} \times 100 = -16.22\% \quad MAC$$

In reality, it is easier to carry out the following calculation on a calculator.

2.5 -1.9 = 0.6
0.6 ÷ 3.7 = 0.162162
0.162162 X 100 = 16.22%

Any value that is below zero percent MAC is forward of the LeMAC. Any value above 100% MAC is further aft than TeMAC.

CONVERTING A PERCENTAGE OF MAC INTO A LINEAR DISTANCE FROM THE DATUM

Where MAC is used, it is standard to give the CG as a percentage of the MAC. However, for calculations requiring the movement of loads, etc., the CG has to be converted back into a linear distance to be a lever arm of the same unit of measure.

For example:

An aircraft has a MAC of 7 ft with the LeMAC located 29 ft aft of the datum. The current CG is located at 26.5% MAC. Calculate to one place decimal the CG's linear distance from the datum.

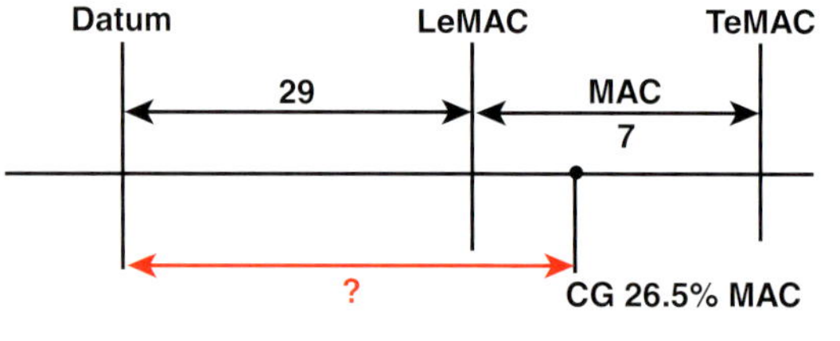

A diagram, which is not to scale, helps to visualise the problem which in this case is to determine the distance of ?. Note that all values are given in feet.

1. Find what 1% of MAC is as a linear value by dividing the linear distance by 100.
2. Multiply this by 26.5%.
3. Add the result to the 29 to get the total dimension.
4. Round the answer to 1pd.

1. 7 ÷ 100 = 0.07 ft per 1%
2. 0.07 ft X 26.5 % = 1.855 ft
3. 1.855 ft + 29 ft = 30.855 ft
4. 30.9 ft - the CG is located 30.9 ft aft of the datum.

CG LIMITS AS PERCENTAGES

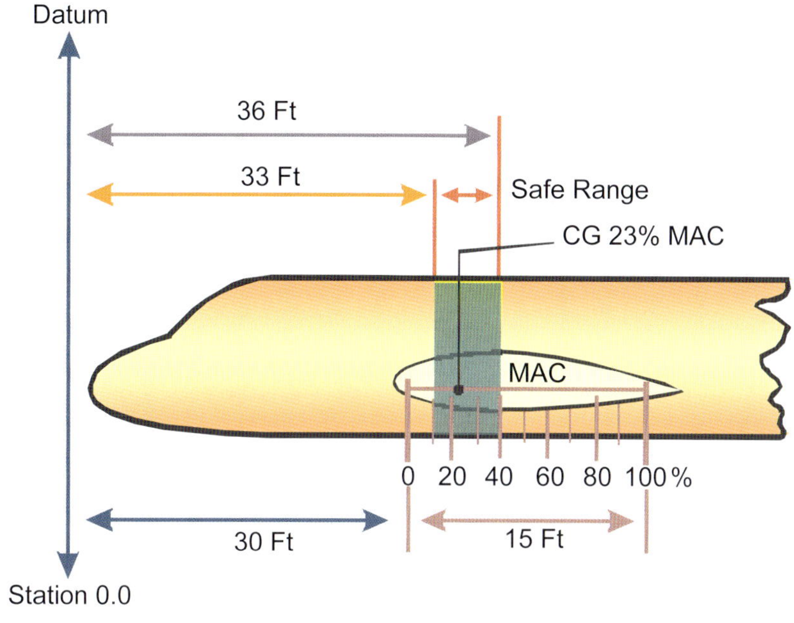

In the diagram above, the fore and aft CG limits can be seen and would normally be given as a percentage of MAC, in this case 10% and 40%. The CG is shown as 23% MAC.

Some exam questions require the conversion of a percentage of MAC into a linear distance and can refer to a line diagram either adjacent to the question or given as an annex at the back of the question paper. For example:

Refer to the diagram above. Express the CG as a linear distance from the datum.

a. 33 ft
b. 33.5 ft
c. 34 ft
d. 34.5 ft

Solution

1. 15 ft ÷ 100% = 0.15 ft
2. 0.15 ft X 23% = 3.45 ft
3. 30 ft + 3.45 ft = 33.45 ft
4. Given answer b. 33.5 ft

In such questions, the minimum of information is provided. Other forms in which such questions are formatted give a set of conditions such as:

An aircraft at TOM of 234 000 kg and CG of 23% MAC will have an estimated fuel burn of 11 000 kg over a four hour flight. The fuel tank arm is 675 in aft of the datum, LeMAC is 613 in aft of the datum and MAC is 139 in in length. Calculate the Gross Mass and CG position at the mid point.

Draw a rough sketch to visualise the problem.

1. Find the CG as a BA from the datum.
2. Determine which BA is the longer, CG or fuel. If it is fuel, the CG moves fwd, and vice versa.
3. Find the total moment of aircraft at TOM.
4. Find the moment for fuel burn at midpoint.
5. Subtract fuel burn moment from TOM moment to find midpoint moment.
6. Subtract fuel burn mass from TOM to find Gross Mass.
7. Find the new CG position as BA.
8. Convert BA into percentage MAC.

1.

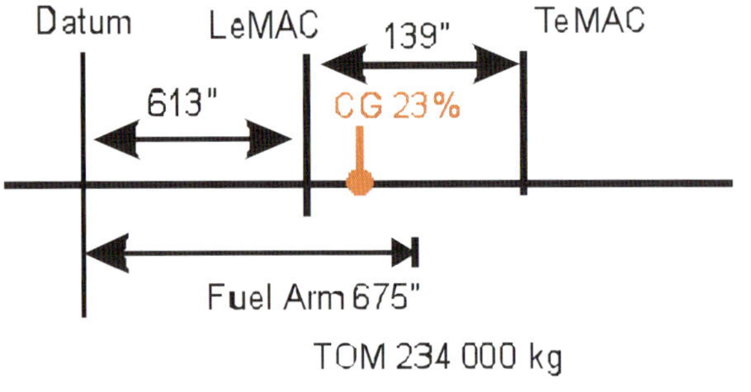

2. 139 in ÷ 100 = 1.39 in per %
   1.39 X 23 = 31.97 in
   613 in + 31.97 in = 644.97 in

3. CG fwd of Fuel BA, so CG will move fwd

4. 644.97 in X 234 000 kg = 150 922 980 kg in Total Moment

5. Fuel Burn at mid point 11 000 kg ÷ 2 = 5500 kg. 5500 kg X 675 in = 3 712 500 kg in

6. 150 922 980 kg in – 3 712 500 kg in = 147 210 480 kg in Gross Mass Moment

7. 234 000 kg - 5500 kg = 228 500 kg Gross Mass

8. 147 210 480 kg in ÷ 228 500 kg = 644.25 in aft of datum

9. $$\frac{644.25 - 613}{139} \times 100 = 22.5\% \text{ MAC}$$

At midpoint, Gross Mass is 228 500 kg with the CG at 22.5% MAC.

In some cases such as fuel consumption, a percentage change can be given for a pre-determined mass of fuel. In these cases, the percentage change can be added or deducted from the original CG as a % MAC as required to find the new CG % MAC. This can also apply to the extending and retracting of flaps.

PRACTICE

1. Choose the correct statement for an aircraft's CG where the LeMAC is + 82 in, the MAC is 60 in, and the CG is located at 20% MAC.

    a. 100 in aft of the datum
    b. 50 in forward of the TeMAC
    c. 12 in aft of LeMAC
    d. 96 in aft of the datum

2. The CG of an aircraft is 22.5 ft aft of datum. MAC is 7 ft, the distance of the trailing edge of the MAC is 29 ft aft of the datum; the CG as a % MAC is?

    a. 7.14%
    b. 17.4%
    c. 27%
    d. 71%

3. A loaded aircraft's CG is located at 24% of an 8 ft MAC, the distance of the leading edge of the MAC is located at + 20 ft, the safe range is from 21% to 36% MAC. Calculate the distance from the datum to the rear CG limit.

    a. 21.92 ft
    b. 21.68 ft
    c. 22.88 ft
    d. 22.92 ft

ANSWERS TO PRACTICE

Question 1

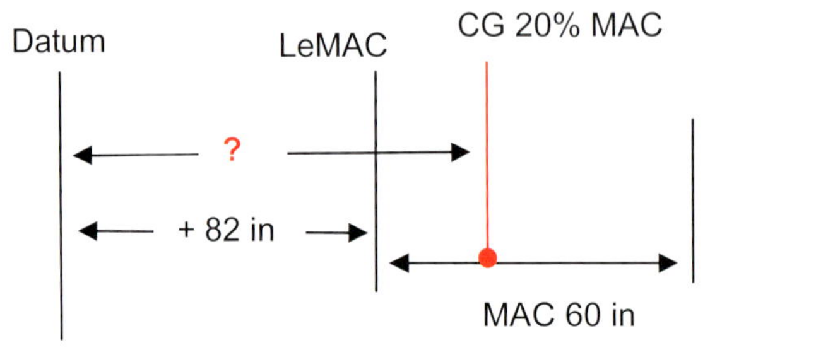

MAC =	60	in
1 % of MAC =	0.6	in
CG at 20% MAC =	12	in
CG in relation to LeMAC =	12	in aft
CG in relation to TeMAC =	48	in fwd
CG in relation to Datum =	94	in aft

Answer C

Question 2

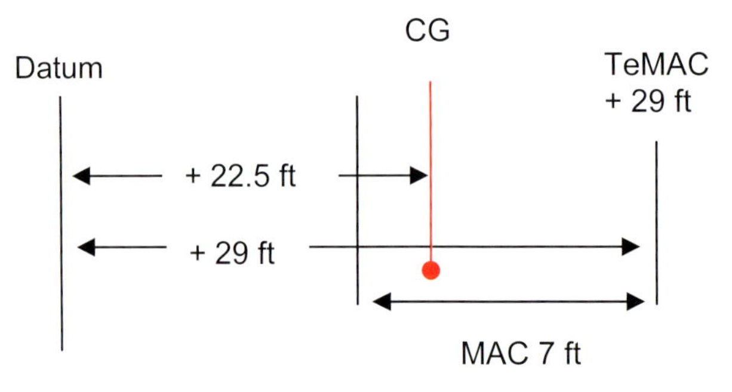

TeMAC to Datum =	29	ft
MAC =	7	ft
LeMAC to Datum =	22	ft
CG to Datum =	22.5	ft
LeMAC to CG =	0.5	ft
CG as a % MAC =	**7.14**	**%**

Answer A

Question 3

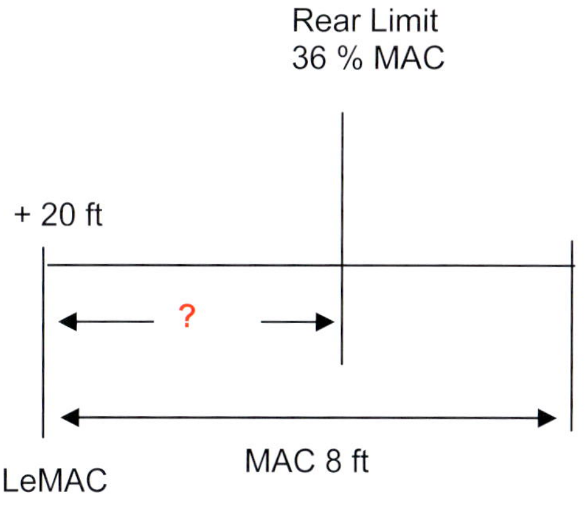

MAC =	8	ft
1% of MAC =	0.08	ft
Rear Limit 36% MAC =	2.88	ft
Datum to LeMAC =	20	ft
Rear Limit to Datum =	**22.88**	**ft**

Answer C

Chapter 8
JAR-Ops 1 Requirements

INTRODUCTION

Subpart J of JAR–Ops 1 details mass and balance requirements for operators of commercial and air transport aircraft. This section of the notes explains the orders and annexes given in this document.

LOADING, MASS, AND BALANCE

The operator of the aircraft is ultimately responsible for loading and the mass and balance of the aircraft and has to specify in the Operations Manual (a company manual) how the aircraft is to be loaded and the method(s) used in calculating both the mass and the CG position. If the aircraft can be operated in more than one role, the method used must cover all of the operations.

MASS AND BALANCE DOCUMENTATION

Mass and balance documentation normally takes the form of either a load manifest and CG envelope or a load and trim sheet. Mass and balance documentation must contain the following:

- The aircraft's registration and type
- The flight identification number and date
- The identity of the Commander
- The identity of the person who prepared the document
- The Dry Operating Mass and the corresponding CG
- The mass of fuel at take-off and the mass of trip fuel
- The mass of consumables other than fuel
- The components of the load including passengers, baggage, freight, and ballast
- The Take-Off Mass, Landing Mass, and Zero Fuel Mass
- The load distribution
- The applicable aeroplane CG position
- The Limiting Mass and CG values

The mass and balance document must enable the Commander of the aircraft to determine the:

- Load
- Distribution of the load
- CG and mass limitations have not been exceeded

The operator of the aircraft is responsible for compiling a mass and balance document prior to each flight. This document must specify what load the aircraft is carrying and how this load is distributed within the aircraft. The person who prepares the document must be named on it, and the person who supervises the actual loading of the aircraft must sign the document to confirm that the aircraft has been loaded accordingly.

The document must be acceptable to the Commander, who certifies acceptance with a signature or with a Personal Identification Number (PIN), if an electronic system is used. The mass and balance document is a legal document, and the PIN has the same legal standing as a signature in certifying acceptance.

LAST-MINUTE CHANGES (LMCS)

Any last-minute changes that occur after the mass and balance document has been completed and the CG calculated for the TOM, LM, and ZFM must be brought to the attention of the Commander and entered on the mass and balance documentation. The operator must specify in the Operations Manual the method by which any Last-Minute Changes (LMCs) are carried out. This limits the number of passengers that can either embark or disembark and the amount of hold load that can be added or removed. If the LMC exceeds this value, new mass and balance documentation must be prepared. If the LMC is within the Operations Manual limit, while the change must be brought to the attention of the Commander, a revised CG does not have to be calculated.

Formerly, flight crews were briefed on the mass and balance aspects of their aeroplane and completed the mass and balance documentation before entering the flight deck. The original copy was stored on the ground and the commander took a copy on board for use in flight.

Modern mass and balance systems are computerised, and some aircraft have a data link between the loading cell (the ground staff involved in the production of m+b docs) and an on-board computer system that allows the crew to self-brief on the flight deck. If this system is used, a hard copy of the accepted document must be available on the ground. To ensure that data links are operating correctly, they must be checked at intervals not exceeding six months.

MASS VALUES FOR CREWS

Operators are required to account for the mass of their crewmembers and baggage as part of determining the dry operating mass. The operator can use:

- The actual mass of the crewmembers and their baggage
- The standard mass of 85 kg for flight crew and 75 kg for cabin crew (regardless of gender or build)
- Any other standard mass acceptable to the Aviation Authority

The operator must correct the DOM and its CG position if the crew has additional baggage.

MASS VALUES FOR PASSENGERS AND BAGGAGE

An operator is required to calculate the mass of passengers (PAX) and their baggage using either actual PAX and baggage masses or by using standard masses given in three tables.

Note: These tables are not printed in the CAP 696 document, so they must be remembered for the examination.

PAX ACTUAL MASSES

If an operator is using actual masses for PAX and baggage, the weighing must occur just prior to boarding and is carried out in the immediate vicinity. The operator must ensure that all the passenger's personal belongings and hand baggage is included in the weighing.

PAX STANDARD MASS CLASSIFICATION

Passengers:

- Under 2 years old are classed as Infants
- Over 2 years but under 12 years are classed as Children
- Over 12 years of age are classed as Adults

If an infant travels on an adult's lap, the infant is determined not to have any accountable mass. However, if the infant travels in its own seat, it is determined as having an accountable mass of 35 kg.

A child of either gender has to travel on its own seat and is determined as having an accountable mass of 35 kg.

Where adult passenger standard masses are used, the age and gender of the passengers is taken into account. These are given in tables 1 and 2 shown below.

TABLE 1 — PASSENGER MASS VALUES FOR UP TO 20 OR MORE SEATS

	Table 1		
Passenger Seats	20 and more		30 and more
	Male	Female	All adult
All flights except holiday charters	88 kg	70 kg	84 kg
Holiday charters	83 kg	69 kg	76 kg
Children	35 kg	35 kg	35 kg

Where the total number of passenger seats in an aeroplane is between 20 and 29, the operator is to use the gender-based masses for 20 and more, as per table 1, for determination of the passenger load. If the aircraft has 30 or more seats, the operator is to use the all adult masses.

All flights except holiday charters	Schedule service, etc.
Holiday charters	Part of holiday package travel
All adult	Adult mass regardless of gender

TABLE 2 — PASSENGER MASS VALUES FOR UP TO 19 SEATS

Table 2			
Passenger seats	**1 – 5**	**6 – 9**	**10 – 19**
Male	104 kg	96 kg	92 kg
Female	86 kg	78 kg	74 kg
Children	35 kg	35 kg	35 kg

For aircraft with 19 passenger seats or less, table 2 applies (see above). The smaller the seating capacity, the greater the standard adult passenger's mass becomes. Within these masses is an allowance for hand baggage. On flights (table 2 only) where no hand baggage is carried in the cabin or where the hand baggage is accounted for separately, 6 kg may be deducted from the adult mass. Articles such as small cameras, overcoats, umbrellas, and small handbags are not considered hand baggage.

TABLE 3 — PASSENGER BAGGAGE VALUES

Table 3 – 20 or more passenger seats	
Type of flight	**Baggage standard mass**
Domestic	11 kg
Within the European region	13 kg
Intercontinental	15 kg
All other	13 kg

Where the total number of passenger seats available is 19 or less, all the passenger hold baggage must be weighed individually to determine the actual baggage mass. Where the pax seats available are 20 or more, standard weights can be applied to each individual item of luggage, as per table 3 shown above.

If an operator wishes to use a different mass as the Company's standard mass, the operator must obtain approval from the Aviation Authorities to conduct a detailed survey. After the authority has approved the results of the survey, only that operator may use differing standard masses and for those types of flights on which the survey was carried out.

If an operator identifies for any flight that a significant number of passengers (including hand baggage) appears to exceed the standard masses used, the operator must determine the actual mass of the these passengers by weighing or adding an adequate mass increment.

If an operator identifies in the hold baggage that there are a significant number of items that could exceed the standard mass used, the operator must determine the actual weight of the luggage by weighing or adding an adequate mass increment.

If the operator has used a non-standard method for determining the mass of passengers or hold baggage, this must be entered into the mass and balance documentation and the commander must be informed.

Domestic is defined as a flight with origin and destination within the border of one state.

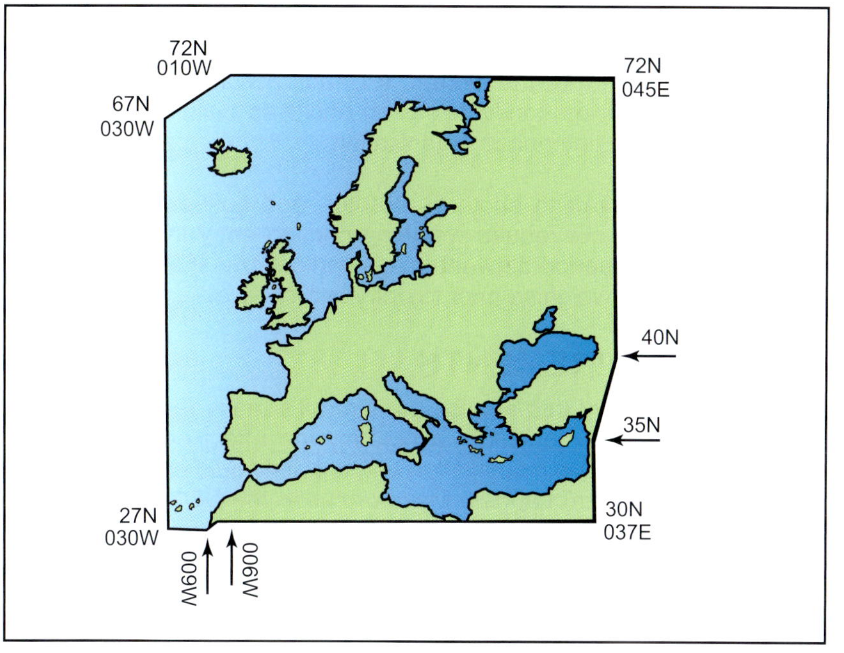

Figure 3.6

Within the European Region is defined as flights being made outside the domestic state, but remaining within the area as defined in the diagram above. The boundary grid is given in table form below.

Intercontinental flights are defined as flights other than within the European Region, meaning flights with their origin and destination in different continents.

N7200 E04500	N3000 W00600	N7200 W01000
N4000 E04500	N2700 W00900	N7200 E04500
N3500 E03700	N2700 W03000	
N3000 E03700	N6700 W03000	

MASS EVALUATION OF AIRCRAFT

When aircraft are manufactured, the manufacturer produces a weighing schedule. This specifies the location of the aircraft's datum and the balance arms to specific items of equipment by detailing their locations and distance from the datum. The datum is found by measuring a specific distance from a physical item on the aeroplane. This item is termed the reference point.

This weighing determines the aircraft's Basic Empty Mass, and its CG is entered on the weighing report, which forms part of the aircraft's technical log.

JAR-Ops 1 requires operators to establish the mass and CG of an aircraft before it enters the operator's service by actual weighing. Provided that the mass and balance documentation has been adjusted for any modification, the factory weighing is counted as this mass and balance evaluation.

For any weighing, the operator must ensure that the aircraft meets the specific conditions as laid down in the weighing schedule for BEM evaluation, that the aircraft is complete, that the fluid levels have been checked, and that the aircraft is clean. This is adjusted to give the standard DOM value. The weighing must be conducted in an enclosed building and is carried out by the manufacturer or an approved maintenance organisation.

If a JAA operator sells an aircraft to another operator, and both have approved mass control programmes, the aircraft does not require weighing prior to entry into the new owner's service provided it is within the 4-year period between mass evaluations. If an aircraft is purchased from a non-JAA operator, it requires weighing prior to entry into service.

RE-WEIGHING REQUIREMENTS

After the initial weighing for entry into service, an aircraft's mass must be re-evaluated every 4 years. Operators are required to document accurately the effect on the mass and CG of any modification or repairs. If cumulative modification and repairs would cause the dry operating mass to alter by ± 0.5% of the Maximum Landing Mass or cause the DOM CG position as given as a percentage of MAC to move by 0.5% in either direction, the aircraft must be reweighed. If the actual effects of modifications or repairs are not accurately known, the aircraft must be weighed.

For example, if an aircraft has a DOM of 1000 kg and a MLM of 18 000 kg with the CG located at 27% MAC, the aircraft would have to be reweighed if the cumulative change to DOM is ± 90 kg or the CG falls on or below 26.5% MAC or on and above 27.5% MAC.

> Take MLM and divide by 100 to find 1% of MLM. Divide this by 2 to find 0.5% MLM.
> Apply this value to the DOM to find the upper and lower tolerance.

For larger operators, a system termed the **Fleet Mass** has been devised. If this system is used, a calendar backstop for weighing individual aircraft is 9 years.

FLEET MASS

The fleet mass system allows operators who have larger fleets of the same model and type of aircraft to simplify their mass and balance documentation. To operate a fleet mass system, the following regulations must be met.

To determine if an aircraft can be placed in a fleet, the operator has to carry out a mass evaluation to determine the fleet's DOM. This is done by weighing a representative proportion of the aircraft that are eligible to be considered for the fleet.

Fleet Mass Determination	
Number of aircraft in the fleet	Minimum number of aircraft to be weighed
2 or 3	n
4 to 9	n + 3 ÷ 2
10 or more	n + 51÷ 10

The number of aircraft to be weighed is determined by the formula given in the table above; **n** is the number of aircraft in the fleet. For example, if an operator has 19 B747 SP aircraft, the operator would have to weigh 7 aircraft to determine the fleet DOM (19 + 51) ÷ 10).

Having established the fleet DOM, a tolerance ±0.5% of the Maximum Landing Mass for the type is applied. This is termed the **fleet mean**. A CG position for the fleet mean is determined as a percentage of MAC and is given a tolerance of ± 0.5% MAC.

The criteria for individual aircraft to enter and remain in the fleet are:

- That its individual DOM is within the fleet mean
- That the individual aircraft's CG position is within the fleet mean

There are exceptions. These are:

1. If an aircraft's DOM is within the tolerance but the CG falls outside the fleet position, the aircraft can continue as part of the fleet but use its own individual CG position.

2. If an individual aircraft has a physical difference, such as galley or toilets located in a different position, seating configuration compared to the rest of the fleet that would accurately account for the mass or CG exceeding the fleet tolerances, the aircraft may remain in the fleet provided corrections are applied to the mass and/or CG for that aeroplane.

Where aeroplanes do not have a mean aerodynamic chord published, they must use individual masses, or the operator must make a special study and submit it to the authorities for approval.

Fleet mass evaluations are carried out to maintain the fleet mean. The maximum time interval between 2 fleet mass evaluations is 48 months and the aircraft selected should be those that have not been weighed for the longest period.

Chapter 9
Loading Manifests, SEP 1, and MEP 1

INTRODUCTION

This chapter details how loading manifests are completed by using the SEP 1 as an example. The process draws attention to the SEP 1 data sheets. The chapter also makes reference to MEP 1. On completion of the chapter, there is a self test paper for the SEP 1 and MEP 1 as both aircraft can be loaded using actual masses or using standard masses from Table 2 of JAR-Ops 1 subpart J. There are a mixture of loads for these questions. While using the loading manifests as given, there are not any references to OM or DOM. These have been included within the test papers to check your retention of previously covered material.

SEP DATA SHEETS — CAP 696

Page 5 of the CAP 696 gives the leading particulars for the SEP 1 generic aircraft. Figure 2.1 shows the reference point (firewall) Datum and the front and rear CG limits. The front CG limit has the typical cut back due to increased stability resulting from increasing mass.

As the aircraft is very light and robust in its construction, the MLM is equal to the MTOM. BEM is the starting point for any mass and balance calculation, since it is a light aircraft. The BEM moment as given in the data sheet is indexed by 100. To prove this, multiply 2415 lb by 77.7 in.

The floor-loading limit is given as an intensity load on page 5 and shown diagrammatically on page 6 as fig. 2.2.

Figure 2.3 at the bottom of page 6 gives the fuel tank arm as 75 in aft of the datum and gives the conversion of gallons to mass and the associated moment indexed by 100.

The gallons are US gallons. As each gallon has a mass of 6 lb (30 lb ÷ 5 gal = 6 lb), the moments given have been rounded when indexed. For example, 30 lb x 75 in = 2250 lb in. This is given as 23. Therefore, it is important to use the table for fuel values when using the SEP 1 data sheet.

A typical Loading Manifest as shown on page 8 fig. 2.4 has been completed for an aircraft with a BEM as per the data sheet with the following load:

Pilot of 140 lb
200 lb of cargo in Zone A
Passengers with a total mass of 120 lb in seats 4 & 5
100 lb in baggage Zone C
Fuel load of 50 US gal, of which 40 US gal will be burnt as trip fuel
Standard starting allowance is made.

In many JAA questions, the same information is presented in the following format. A typical Loading Manifest as shown on page 8 fig. 2.4 has been completed for an aircraft with a BEM as per the data sheet with the following load:

Pilot of 140 lb, 200 lb of cargo in Zone A, passengers with a total mass of 120 lb in seats 4 & 5, 100 lb in baggage Zone C, a fuel load of 50 US gal, of which 40 US gal will be burnt as trip fuel. A standard starting allowance is made.

Column 1	2	3	4
ITEM	**MASS** (lb)	**ARM (in)**	**MOMENT X 1000**
1. Basic Empty Condition	**2415**	**77.7**	**1876.46**
2. Front Seat Occupants	140	79	110.6
3. Third & Fourth Seat Pax	0	117	0
4. Baggage Zone A	200	108	216
5. Fifth & Sixth Seat Pax	120	152	182.4
6. Baggage Zone B	0	150	0
7. Baggage Zone C	100	180	180
Sub–Total = ZERO FUEL MASS	**2975**	**86.2**	**2565.46**
8. FUEL Loading	300	75	225
Sub–Total = RAMP MASS	**3275**	**85.2**	**2790.46**
9. Subtract Fuel for Start, Taxi, and Run Up (see Note)	**-13**		**-10**
Sub–Total = TAKE-OFF MASS	**3262**	**85.2**	**2780.46**
10. TRIP FUEL	**- 240**		**-180**
Sub–Total = LANDING MASS	**3022**	**86.4**	**2610.46**

The manifest lays out in a tabular and logical order the information about the aircraft and the intended load to be carried, allowing the calculation of mass and CG for the three primary conditions (ZFM, TOM, and LM). Listed below is a breakdown of how the example manifest is compiled.

Line 1 The information for the BEM, CG, and moment are taken from page 5 of the data sheets. The BEM entered in column 2, the CG is entered in column 3, and the moment in its indexed form in column 4. These are shown in green in the example.

Line 2 The mass of the front seat occupants (a pilot in this case) is entered in column 2 and multiplied by the arm printed in column 3. The result is divided by 100 and entered into column 4.

Line 3/4 The aircraft can either carry passengers (Pax) or baggage in this area. If Pax are carried, the mass is entered on line 3. If baggage is carried instead of Pax, the mass is entered on line 4, and the moment is calculated using the baggage compartment centroid. In this case, the mass is baggage and is entered on line 4.

Line 5/6 As above, the aircraft can either carry passengers (Pax) or baggage in this area. If Pax are carried, the mass is entered on line 5. If baggage is carried instead of Pax, the mass is entered on line 6, and the moment is calculated using the Pax seat centroid. In this case, the mass is Pax and is entered on line 5. Care must be taken entering the mass in the appropriate line as the balance arms are of different lengths. Care must also be taken to ensure that any item does not exceed the load intensity for the floor if baggage is loaded. This does not show in the manifest.

Line 7 This is a dedicated baggage compartment and has the last printed arm.

Sub-Total = Zero Fuel Mass
For column 2, calculate the ZFM by totalling all the masses entered in lines 1 to 7 and enter the value in column 2. For column 4, calculate the ZFM moment by totaling the moments in column 4. Find the ZFM CG by dividing the ZFM moment by the mass. Enter this value in column 3. Check that this value is within the MZFM if one applies, and that the CG is within the CG limits.

Line 8 If the ZFM and CG are within limits, the mass of the fuel load is added in line 8. The table, fig. 2.3 on page 6, converts the US gallons into pounds and gives the moment as an indexed value. These values are entered in columns 2 and 4.

As the fuel load in the table is given in 5 gallon increments, if the fuel load given is between those of the table, calculate the mass and the moment. Use the constant of 6 lb per gallon and the arm as + 75 in.

For example:
7.5 gal have a mass of 45 lb and have an indexed moment of 34 lb in.

7.5 gal X 6 lb = 45 lb
45 lb X 75 in = 3375 lb in
3375 lb in ÷ 100 = 33.75 lb
Indexed to a full number= 34 lb in

A cross check for 7.5 gallons using the table:

60 lb - 30 lb = 30 lb, 30 lb ÷ 2 = 15 lb, 15 lb + 30 lb = 45 lb
45 lb in – 23 lb in = 22 lb in, 22 lb in ÷ 2 = 11 lb in, 11 lb in + 23 lb in = 34 lb in

Sub–Total = Ramp Mass
For column 2, calculate the Ramp Mass by totaling the ZFM and the fuel load masses and enter the value in column 2. For column 4, total the ZFM moment and the fuel moments in column 4.

Find the ramp mass CG by dividing the ramp mass moment by the ramp mass. Enter this value in column 3. This is not required for the SEP 1 but is required for aircraft where there is a maximum structural ramp mass above the maximum take-off mass.

Line 9 Insert the value of -13 lb in the mass column. This is the value given below the load manifest in fig. 2.4 page 8. Enter the value -10 in column 4. Placing the minus symbol in front of the values ensures that they are not added by mistake.

Sub–Total = Take-Off Mass
Subtracting the start allowance from the ramp mass and the start fuel moment from the ramp moment gives the take-off mass values from which the take-off CG can be calculated.

Line 10 If the take–off CG and mass are within limits, enter the trip fuel mass and moment as negatives in columns 2 and 4.

Sub–Total = Landing Mass
Subtracting the trip fuel mass and moment from the take–off mass and moment gives the landing mass and moment from which the landing CG can be calculated.

If the ZFM, TOM, and LM masses and CG positions are in limits, the aircraft is safe to fly. If the mass is within limits but the CG is aft of the rear limit, either a rearward mass must be removed to locate the CG on or forward of the rear limit or relocated forward so as to shift the CG on or forward of the rear limit. Conversely, if the CG is forward of the front limit, a forward mass must either be relocated rearward or removed to relocate the CG on or behind the front limit.

MEP 1

Column 1	2	3	4
ITEM	**MASS** (lb)	**ARM Aft Of Datum (in)**	**MOMENT (in/lb)**
Basic Empty Mass	**3210**	**88.5**	
Pilot and Front Passenger		85.5	
Passengers (centre seats) or Baggage Zone 2 (360 lb Max)		118.5	
Passengers (rear seats) or Baggage Zone 3 (400 lb Max)		157.6	
Baggage Zone 1 (100 lb)		22.5	
Baggage Zone 1 (100 lb)		178.7	
ZERO FUEL MASS (4470 lb Max-Std)			
Fuel (123 gal Max).		**93.6**	
RAMP MASS (4773 lb Max)			
Fuel allowance for Start, Taxi, Run-Up		**93.6**	
TAKE-OFF MASS (4750 lb Max)			
Minus Estimated Fuel Burn-Off		**93.6**	
LANDING MASS (4513 lb Max)			

The table above is a reproduction of the MEP 1's loading manifest as shown on page 16 as figure 3.4 of the CAP 696. Notice that this aircraft has a Maximum ramp mass that is above MTOM. As the aircraft operates at higher masses, there is a structure landing mass.

Completing the load manifest is similar to that of the SEP 1. However, notice that the passenger seating and baggage arms are the same and that all fuel masses and moments have to be calculated. Also note that the moments for each part of the load are entered in their full form.

It is the Commander's responsibility to check that the load manifest is correctly completed and that the aircraft is safe for the flight intended. In training to become a Commander, do not in real life or in any test accept that the values given in the manifest are always 100% correct. In some JAA questions, there are deliberate mistakes, which, if followed, lead to the selection of an incorrect answer that is given as an option to catch the unwary.

For example, refer to the loading manifest figure 3.2 page 14 of CAP 696. Select the correct answer for the landing mass.

a. 4300 lb at + 93.6 in
b. 4298 lb at + 93.9 in
c. 4300 lb at - 93.6 in
d. 4298 lb at - 93.9 in

Working shown in table below:

Item	Mass	Arm	Moment
BEM	3210	88.5	284 085
Front	340	85.5	29 070
Centre	236	118.5	27 966
Rear	340	157.6	53 584
Zone 1	100	22.5	2250
Zone 2	0	178.7	0
ZFM	**4226**	**93.9**	**396 955**
Fuel	545	93.6	51012
Ramp	**4771**	**93.9**	**447 967**
Start	-23	93.6	-2152.8
TOF	**4748**	**93.9**	**445 814.2**
Trip	-450	93.6	-42120
LM	**4298**	**93.9**	**403 694.2**

QUESTIONNAIRE FOR DATA SHEET SEP 1

Use the conversion factors given in the CAP 696 page 4, reduced to 4 places decimal.

Work with the converted values, then give the final answer for CG and moments to two places decimal and convert masses to nearest whole number.

1. What is the MTOM of the aircraft?

2. What is the maximum load per square foot in Zone C?

3. What distance is the datum from the Reference Point?

4. What is the mass of a single gallon of fuel?

5. What is the moment of a single gallon of fuel?

6. What is the normal run-up and taxi allowance of fuel?

7. What is the DOM + CG for a crew of two pilots each weighing 182 lb?

8. What is the aircraft's gross mass and CG if it is parked with full fuel tanks?

9. What is the OM for an aircraft crewed with two pilots each weighing 182 lb and fully fueled?

10. An aircraft is operated with six seats, one crewmember, and five passengers using the standard masses where each passenger is a male adult without hand baggage. Two hundred pounds of baggage is loaded into Zone C. What would its ZFM and CG be?

11. What would be the TOM, LM, and ZFM if the aircraft were to be operated in the two-seat configuration as loaded below for a six-hour flight?

    - 1 Crew (standard mass)
    - 1 Adult female (standard mass) in front seat
    - Full fuel load with a fuel burn of 5 gallons per hour

12. A small ad hoc charter company operates an aircraft with a male and female crew who weigh 189 lb and 104 lb respectively. The aircraft is chartered to fly cargo between two airports, 2½ hours apart. The company operations manual states that a contingency reserve must be carried for each operation that is the greatest of either 10% of the flight fuel or 3 gallons. The fuel consumption is estimated to be 6 gal per hour. Calculate the masses and determine the maximum payload that could be carried allowing a standard start fuel allowance.

13. An aircraft is operated with only the pilots seats fitted, fully fuelled with two female pilots at standard mass. What floor area would be required to support a payload of 416 lb loaded into area A?

14. An aircraft is to be loaded as below. Calculate the aircraft's useful load.

    Front seats - pilot weighs 112 lb, pax weighs 196 lb.
    The 3+4 seats - female weighs 105 lb and child weighs 56 lb.
    Baggage in Zone B = 200 lb
    Baggage in Zone C = 100 lb
    Fuel load 360 lb
    Trip fuel 180 lb
    Start fuel 13 lb

	Mass	Arm	Moment (X 100)
BEM	2415.00	77.7	
Front seats		79	
3+4 seats		117	
Baggage Zone A		108	
5+6 seats		152	
Baggage Zone B		150	
Baggage Zone C		180	
ZFM			
Fuel Load			
Ramp mass			
Less Start			
Take-Off Mass			
Trip			
Landing Mass			

Work out the CG and mass for the following conditions:

10. DOM, OM, ZFM, Taxi Mass, TOM, and LM for an aircraft loaded as follows:

    Pilot 160 lb
    Pax 238 lb in seats 3+4
    Pax 126 lb in seats 5+6
    Baggage 100 lb in Zone C
    Fuel load is 25 gal, including reserve and standard start allowance.
    Fuel burn is 6.5 gal per hour.
    Trip time is 2.5 hours.

	Mass	Arm	Moment (X 100)
BEM	2415.00	77.7	
Front seats		79	
3+4 seats		117	
Baggage Zone A		108	
5+6 seats		152	
Baggage Zone B		150	
Baggage Zone C		180	
ZFM			
Fuel Load			
Ramp mass			
Less Start			
Take-Off Mass			
Trip			
Landing Mass			

QUESTIONNAIRE FOR DATA SHEET MEP 1

1. What is the maximum load for Zone 2?

2. What is the MTOM?

3. What is the MZFM?

4. If standard passenger masses from Table 2 of JAR-OPS 1 subpart J are used for this aircraft, what is the moment for a female adult and male child sitting in the rear row of seats?

5. Calculate the mass of the aircraft given an arm of 89.5 in aft of the datum and a moment of 416 175 in lb and with reference to the CG limits.

6. What mass is allowed for each US gallon of fuel?

7. Where is the reference point for the aircraft?

8. What distance is the nose wheel from the main wheel?

9. What effect does raising the undercarriage have?

10. Why does the forward CG limit decrease from 82 in to 91 in?

11. What is the performance class of this aeroplane?

12. What is the MLM?

13. What is the difference in mass for the specimen aircraft at a gross mass of BEM and full fuel and the MTOM?

14. If the CG plot falls on the line it is?

15. What is the fwd limit of the safe range at 4300 lb mass?

16. Calculate the take-off mass, landing mass, and CG positions for a specimen aircraft operated as a freighter, which has the following load:

    100 lb Zone 1
    360 lb Zone 2
    400 lb Zone 3
    100 lb Zone 4
    Crew 183 lb
    Bulk fuel 21 US gal
    Trip fuel 15 US gal
    Start fuel 16 lb

    a. TOM in limits take-off CG out of limits, landing mass and CG out of limits
    b. TOM out of limits take-off CG in limits, landing mass and CG in of limits
    c. All limits exceeded
    d. All masses within limits, CG limits exceeded

17. For the specimen aircraft loaded as below, calculate the following masses and CG, DOM, ZFM, TOM, and LM.

ITEM	**Mass (lb)**	Arm Aft Of Datum (in)	MOMENT (in lb)
Basic Empty Mass	3210	88.5	
Pilot and Front Passenger	182	85.5	
Passengers (Centre Seats) or Baggage Zone 2 (360 lb Max)	294	118.5	
Passengers (Rear Seats) or Baggage Zone 3 (400 lb Max)		157.6	
Baggage Zone 1 (100 lb Max)	60	22.5	
Baggage Zone 4 (100 lb Max)	100	178.7	
Zero Fuel Mass (4470 lb Max - Std)			
Fuel (123 gal Max) **79 gal**			
Ramp Mass (4773 lb Max)			
Start Fuel	-30		
Take-Off Mass (4750 lb Max).			
Minus trip fuel **32 gal**			
Landing Mass (4513 lb Max).			

18. A specimen aircraft where Zone 3 is used as a freight bay is chartered to transport two pax and a package from Airport A to Airport B, a distance of 350 nm. Airport B is unable to refuel the aircraft. The fuel load at TOM from airport A must include all the fuel required for start, trip, and reserve.

    Flight 360 nm at 120 kt
    Fuel consumption is 10 US gal per hour outbound and 7.5 US gals per hour inbound
    A start taxi allowance for the departure airport is 16 lb and 10 lb for the start from the destination airport. The aircraft must land with a diversion fuel allowance of 1.5 hour.
    Pilots: Capt. 138 lb
    Training Capt. 203 lb
    Freight Baggage Zone 2: Freight marked 160 kg Dimension 1 m X 0.5 m X 0.5 m
    Freight Baggage Zone 3: Freight marked 83 lb Dimension 2.75 ft X 1.5 ft X 0.5 ft

    a. The aircraft can make both the outbound and return flight
    b. The aircraft is over the limits for the outbound landing
    c. The aircraft is over the limits for the outbound take-off
    d. The aircraft can make the outbound but not return flight

19. An aircraft is loaded as follows:

    Pilot 158 lb
    1st row pax 198 lb
    2nd row pax 126 + 130 lb
    3rd row pax 0 lb
    Bulk fuel 100 US gal
    Baggage 100 lb Zone A and 4
    Trip fuel 73 gal
    Fuel allowance for start -15 lb

    Is the aircraft:

    a. In limits for take-off
    b. In limits for landing
    c. In limits for take-off and landing
    d. Not in limits for either take-off or landing

Below is a load manifest that has been prepared for a planned flight to transport three items of freight and one passenger. Check the manifest and select the correct statement from those listed below.

ITEM	Mass (lb)	Arm Aft of Datum (in)	Moment (in lb)
Basic Empty Mass	3210.00	88.5	284 085.00
Pilot and Front Passenger	340.00	85.5	29 070.00
Passengers (Centre Seats) or			
Baggage Zone 2	360.00	118.5	42 660.00
Passengers (Rear Seats) or			
Baggage Zone 3	240.00	157.6	37 824.00
Baggage Zone 1	100.00	22.5	2250.00
Baggage Zone 4	77.00	178.7	1375.99
Zero Fuel Mass	**4327.00**	**91.8**	**397 264.99**
Fuel	444.00	93.6	41 558.40
Ramp Mass	**4771.00**	**92.0**	**438 823.39**
Start Fuel	-22	93.6	-2059.20
Take-Off Mass	**4749.00**	**92.0**	**436 764.19**
Minus trip fuel	-400.00	93.6	-37 440.00
Landing Mass	**4349.00**	**91.8**	**399 324.19**

a. The aircraft is in limits for ZFM, RM, TOM, and LM.
b. The aircraft is in limits for TOM but not for ZFM and LM.
c. The aircraft is in limits for ZFM and RM but not for TOM and LM.
d. The aircraft is in limits for ZFM, RM TOM but not LM.

ANSWERS FOR QUESTIONNAIRE DATA SHEET SEP1

All page numbers refer to CAP 696.

1. **3650 lb** Page 5.

2. **100 lb per square foot** Page 5

3. **39 inches** Page 5

4. **6 lb** Page 6 30 lb ÷ 5 = 6 lb per gallon

5. **460 lb/in** Page 6 23 lb in ÷ 5 = 4.6 (moment/100)

6. **13 lb** Page 8

7. **DOM = 2779 lb at + 77.87 in**

	Mass	Arm	Moment (X 100)
BEM	2415	77.7	187 645.5
Crew	364	79	28 756
DOM	2779		216 401.5
DOM CG	216 401.5	Divided by	2779
CG =	+ 77.87	Inches	

8. **GM 2859 lb at + 77.28 in**

	Mass	Arm	Moment (X 100)
BEM	2415	77.7	187 645.5
Fuel	444	75	33 300.0
Gross	2859		220 945.5
DOM CG	220 945.5	Divided by	2859
CG =	+ 77.28	Inches	

9. **OM 3223lb**

	Mass
BEM	2415
Crew	364
Fuel	444
OM	3223

10. **ZFM 3883 lb at 95.74 in**

From the standard passenger mass for a 1-5 passenger seat aircraft, a male's mass = 104 kg. This includes 6 kg of hand baggage. Each male in this question has a mass of 98 kg. The crew has a standard mass of 85 kg. Convert the mass from kg to lb by multiplying it by the constant.

Crew 85 kg	X	2.2046 lb	187.39 lb
Pax 98 kg	X	2.2046 lb	216.05

	Mass	Arm	Moment (X 100)
BEM	2415.00	77.7	187 645.50
Front seats	403.44	79	31 871.76
3+4 seats	432.10	117	50 555.70
Baggage Zone A	0	108	0
5+6 seats	432.10	152	65 679.20
Baggage Zone B	0	150	0
Baggage Zone C	200.00	180	36 000.00
ZFM	3882.64		371 752.16

ZFM = 3883 lb
ZFM CG is 371 752.6 lb/in divided by 3883
CG = + 95.74 in

11. **ZFM 2792 lb at 77.88 in**
TOM 3223 lb at 77.76 in
LM 3043 lb at 77.93 in

As a pax is female but hand luggage is not mentioned, assume it is there.

Female 86 kg X 2.2046 = 189.60 lb
Crew 85 kg X 2.2046 = 187.39 lb

Total mass is 376.99 lb. As this is 0.01 less than a pound, round it up for use in the table.

	Mass	Arm	Moment (X 100)
BEM	2415.00	77.7	187 645.50
Front seats	377.00	79	29783
3+4 seats	0.00	117	0.00
Baggage Zone A	0	108	0
5+6 seats	0.00	152	0.00
Baggage Zone B	0	150	0
Baggage Zone C	0.00	180	0.00
ZFM	2792.00	**77.88**	217 428.50
Fuel Load	444.00	75	33 300.00
Ramp mass	3236.00	**77.48**	250 728.50
Less Start	-13		-100.00
Take-Off Mass	3223.00	**77.76**	250 628.5
Trip 6 x 5 x 6	-180.00	75	-13 500.00
Landing Mass	3043.00	**77.93**	237 128.5

12. **Maximum payload 821 lb**

		Mass
BEM		2415.00
Front seats		293.00
3+4 seats		0.00
Baggage Zone A		0
5+6 seats		0.00
Baggage Zone B		0.00
Baggage Zone C		0.00
ZFM		2708.00
Fuel Load		121.00
Ramp mass		2829.00
Less Start		-13
Take-Off Mass		2816.00
Trip 2.5 x 6 x 6		-90.00
Landing Mass		2726.00

Trip fuel is calculated as 2.5 hrs X 6 gal = 15 gallons. The mass for this fuel is found in fig. 2.3 or calculated at 6 lb per gallon, 15 gal X 6 lb = 90 lb. The fuel load has to be increased by the greatest of either 10% of the trip fuel or 3 gallons. Ten percent of 15 gallons is 1.5 gallons; therefore, the fuel reserve to be used is 3 gallons, equaling 18 lb. A start allowance of 13 pounds is also required, giving a total fuel load of 121 lb (90 + 18 + 13).

The data sheet for this aircraft shows that the MTOM 3650 lb is equal to MLM and with reference to the CG envelope there is no extra allowance for ramp mass. Given this, the maximum payload the aircraft can carry is found by subtracting the ramp mass from 3650 lb.

Max payload is: 3650 lb – 2829 lb = 821 lb

This accounts for the start fuel allowance on board the aircraft. If the actual take-off mass of 2816 lb is used, it would give a value of 834 lb for the payload. This would make the aircraft 13 lb overloaded at the ramp.

13. **8.32 sq ft**

This type of question gives unwanted information to catch the unwary. The actual question asked is what square area is required to support a load of 416 lb given the structural limitation for Zone A (50 lb per sq ft).

416 lb ÷ 50 lb/sq ft = 8.32 sq ft

14 **1017 lb Useful load**

The JAA definition of useful load is only given in the theory syllabi and is the combined masses of fuel and payload.

Pax Mass	Baggage	Fuel
196	B 200	Bulk 360
105	C 100	
56		
Sub-Total 357	Sub-Total 300	Sub-Total 360
Total 1017 lb		

15.

DOM	2575.00	at 77.78 in	**Taxi Mass**	3189.00	at 86.71 in
OM	2712.00	at 77.96 in	**TOM**	3176.00	at 87.04 in
ZFM	3039.00	at 87.29 in	**LM**	3078.50	at 87.42 in

		Mass	Arm	Moment (X 100)
BEM		2415.00	77.7	1876.46
Front seats		160.00	79	126.4
3+4 seats		238.00	117	278.46
Baggage Zone A		0.00	108	0
5+6 seats		126.00	152	191.52
Baggage Zone B		0.00	150	0
Baggage Zone C		100.00	180	180
ZFM		3039.00	**87.29**	2652.84
Fuel Load		150.00	75	112.50
Ramp Mass		**3189.00**	**86.71**	2765.34
Start fuel		-13.00	75	-10.00
TOM		**3176.00**	**87.04**	2764.34
Trip (2.5 x 6.5 x 6)		-97.50	75	-73.13
LM		**3078.50**	**87.42**	2691.21

	Mass	Arm	Moment (X 100)
BEM	2415.00	77.7	1876.46
Crew	160.00	79.00	126.4
DOM	**2575.00**	**77.78**	2002.86
TOF	137.00	75.00	102.75
OM	**2712.00**	**77.96**	2114.36

ANSWERS FOR QUESTIONNAIRE DATA SHEET MEP1

Page numbers refer to CAP 696

1. 360 lb Page 13
2. 4750 lb Page 12
3. 4470 lb Page 12
4. 42 041.38 lb in

Standard Pax Mass from Table 2 for 1-5 Pax seats			
	kg	Conversion	Pounds
Female	86	2.2046	189.60
Child	35	2.2046	77.16
Total Mass	121	2.2046	266.76
	Mass	Arm	Moment
Passengers Rear Seats	266.76	157.6	**42 041.38**

5. 416 175 lb in ÷ 89.5 in = 4650 lb which is fwd of the fwd limit for the mass Page 15
6. 6 lb per US gallon Page 13
7. Leading edge of the wing at inboard edge of inboard fuel tank Page 12
8. 84.5 in distance B – A (109.8 in – 25.3 in) Page 12
9. No significant effect Page 12
10. Due to the increase in mass, the aircraft's fwd limit of the CG is reduced to prevent the aircraft becoming too stable and requiring excessive force and excessive deflection to operate the controls.
11. Performance class B Page 12
12. 4513 lb Page 12
13. 802 lb

Basic Empty Mass	3210.00
Fuel (123 gal x 6 lb)	738.00
Gross Mass	3948.00
MTOM	4750.00
Gross Mass	-3948.00
Difference	**802.00**

14. In limits
15. **+87.1 in.** Using the CG envelope on page 15, find mid point between the 4200 and 4400 in two places. Then project a horizontal line across the Fwd safe limit where they intersect. Drop a line down to the CG location scale that parallels the nearest grid line.

16. All masses within limits, CG limits exceeded

ITEM	Mass (lb)	Arm Aft of Datum (in)	Moment (in lb)
Basic Empty Mass	3210.00	88.5	284 085.00
Pilot and Front Passenger	183.00	85.5	15 646.50
Passengers (Centre Seats) or			
Baggage Zone 2 (360 lb Max)	360.00	118.5	42 660.00
Passengers (Rear Seats) or			
Baggage Zone 3 (400 lb Max).	400.00	157.6	63 040.00
Baggage Zone 1 (100 lb Max)	100.00	22.5	2250.00
Baggage Zone 4 (100 lb Max)	100.00	178.7	17 870.00
Zero Fuel Mass (4470 lb Max - Std)	**4353.00**	**97.8**	**425 551.50**
Fuel (123 gal Max)	126.00	93.6	11 793.60
Ramp Mass (4773 lb Max)	**4479.00**	**97.6**	**437 345.10**
Start Fuel	-16	93.6	-1497.60
Take-Off Mass (4750 lb Max).	**4463.00**	**97.7**	**435 847.50**
Minus trip fuel	-90.00	93.6	-8424.00
Landing Mass (4513 lb Max).	**4373.00**	**97.7**	**427 423.50**

17.

DOM 3392 lb at +88.3 in	**TOM 4290 lb at +92.1 in**
ZFM 3846 lb at +92.0 in	**LM 3079 lb at +92.1 in**
RM 4320 lb at +92.1 in	

ITEM	Mass (lb)	Arm Aft of Datum (in)	Moment (in lb)
Basic Empty Mass	3210.00	88.5	284 085.00
Pilot and Front Passenger	182.00	85.5	15 561.00
Passengers (Centre Seats) or			
Baggage Zone 2 (360 lb Max).	294.00	118.5	34 839.00
Passengers (Rear Seats) or			
Baggage Zone 3 (400 lb Max).	0.00	157.6	0.00
Baggage Zone 1 (100 lb Max).	60.00	22.5	1350.00
Baggage Zone 4 (100 lb Max).	100.00	178.7	17 870.00
Zero Fuel Mass (4470 lb Max - Std)	**3846.00**	**92.0**	**353 705.00**
Fuel (123 gal Max)	474.00	93.6	44 366.40
Ramp Mass (4773 lb Max)	**4320.00**	**92.1**	**398 071.40**
Start Fuel	-30	93.6	-2808.00
Take-Off Mass (4750 lb Max).	**4290.00**	**92.1**	**395 263.40**
Minus trip fuel	-192.00	93.6	-17971.20
Landing Mass (4513 lb Max).	**4098.00**	**92.1**	**377 292.20**
BEM	3210.00	88.5	284 085.00
Crew	182.00	85.5	15 561.00
DOM	**3392.00**	**88.3**	**299 646.00**

18. **The aircraft can make both the outbound and return flight**

First check that the freight is within the floor load limit of 120 lb per sq ft (Page 12).

	kg	Conversion	lb		
Freight	160.00	2.2046	352.74		
Item 1 Dimensions 1 m x 0.5 m x 0.5 m					
Metres	Conversion	ft	sq ft	Load lb	lb per sq ft
1 m	0.3048	3.28	5.38	352.74	65.57
0.5 m	0.3048	1.64	2.69	352.74	131.08
0.5 m	0.3048	1.64			
Item 2 Dimensions 2.75 ft x 1.5 ft x 0.5 ft					
		2.75	4.13	83	20.09
		1.5	0.75	83	110.67
		0.5			

Both items are in floor load limits.

Calculate the fuel load for take–off from the departure airport by calculating the flight time, etc. On landing at B, the aircraft has the return trip fuel on board, so the reserve fuel is calculated at the consumption rate for the return flight.

Distance	Speed	Flight Time			
360	120	3.00			
				Gallons	Mass
Trip fuel Out Bound 3 hr at 10 gal at 6 lb				30.00	180.00
Trip fuel Inbound 3 hr at 7.5 gal at 6 lb				22.50	135.00
Total Trip Fuel				52.50	315.00
Start allowance A 16 lb				2.67	16.00
Start allowance B 10 lb				1.67	10.00
Reserve 1.5 x 7.5 gph x 6 lb				11.25	67.50
Fuel Load at A Max 123 gal 738 lb				68.09	408.5

As both the fuel and the freight are in limits, calculate the outbound manifest.

For Flight from A to B	Mass (lb)	Arm (in)	Moment (in lb)
Basic Empty Mass	3210.00	88.5	284 085.00
Pilot and Front Passenger	341.00	85.5	29 155.50
Passengers (Centre Seats) or			
Baggage Zone 2 (360 lb Max).	352.74	118.5	41 799.22
Passengers (Rear Seats) or			
Baggage Zone 3 (400 lb Max).	83.00	157.6	13 080.80
Baggage Zone 1 (100 lb Max).	0.00	22.5	0.00
Baggage Zone 4 (100 lb Max).	0.00	178.7	0.00
Zero Fuel Mass (4470 lb max)	**3986.74**	**92.3**	**368 120.52**
Fuel (123 gal Max)	408.50	93.6	38 235.60
Ramp Mass (4773 lb Max)	**4320.00**	**94.1**	**406 356.12**
Start Fuel	-16.00	93.6	-1497.60
Take-Off Mass (4750 lb Max).	**4304.00**	**94.1**	**404 858.52**
Minus trip fuel	-180.00	93.6	-16 848.00
Landing Mass (4513 lb Max).	**4124.00**	**94.1**	**388 010.52**

Check the outbound actual masses against the structural maximum masses.
Check the calculated CG position against the CG envelope.
As the outbound flight is in limits, calculate the inbound flight.

For Flight from B to A	Mass (lb)	Arm (in)	Moment (in lb)
Basic Empty Mass	3210.00	88.5	284 085.00
Pilot and Front Passenger	341.00	85.5	29 155.50
Passengers (Centre Seats) or			
Baggage Zone 2 (360 lb Max).	0.00	118.5	0.00
Passengers (Rear Seats) or			
Baggage Zone 3 (400 lb Max).	0.00	157.6	0.00
Baggage Zone 1 (100 lb Max).	0.00	22.5	0.00
Baggage Zone 4 (100 lb Max).	0.00	178.7	0.00
Zero Fuel Mass (4470 lb max)	**3551.00**	**88.2**	**313 240.50**
Fuel (123 gal Max)	212.50	93.6	19 890.00
Ramp Mass (4773 lb Max)	**3763.50**	**88.5**	**333 130.50**
Start Fuel	-10.00	93.6	-936.00
Take-Off Mass (4750 lb Max).	**3753.50**	**88.5**	**332 194.50**
Minus trip fuel	-135.00	93.6	-12 636.00
Landing Mass (4513 lb Max).	**3618.50**	**88.3**	**319 558.50**

Check the inbound actual masses against the structural maximum masses.
Check the calculated CG position against the CG envelope.

19. In limit for take-off and landing.

ITEM	Mass (lb)	Arm Aft of Datum (in)	Moment (in lb)
Basic Empty Mass	3210.00	88.5	284 085.00
Pilot and Front Passenger	356.00	85.5	30 438.00
Passengers (Centre Seats) or			
Baggage Zone 2 (360 lb Max).	256.00	118.5	30 336.00
Passengers (Rear Seats) or			
Baggage Zone 3 (400 lb Max).	0.00	157.6	0.00
Baggage Zone 1 (100 lb Max).	100.00	22.5	2250.00
Baggage Zone 4 (100 lb Max).	100.00	178.7	17 870.00
Zero Fuel Mass (4470 lb Max - Std)	**4022.00**	**90.7**	**364 979.00**
Fuel (123 gal Max)	600.00	93.6	56 160.00
Ramp Mass (4773 lb Max)	**4622.00**	**91.1**	**421 139.00**
Start Fuel	-15.00	93.6	-1404.00
Take-Off Mass (4750 lb Max).	**4607.00**	**91.1**	**419 735.00**
Minus trip fuel	-450.00	93.6	-42 120.00
Landing Mass (4513 lb Max).	**4157.00**	**90.8**	**377 615.00**

20. The aircraft is in limits for take-off but not for ZFM and LM.

Purposeful error made in Baggage Zone 4 line (77 X 178.7 = 13 759.9). The JAA uses this type of question to check that the candidate is checking the computation of others rather than accepting a manifest and just signing for it.

ITEM	Mass (lb)	Arm Aft of Datum (in)	Moment (in lb)	True (in lb)	True CG
Basic Empty Mass	3210.00	88.5	284 085.00	284 085	
Pilot and Front Passenger	340.00	85.5	29 070.00	29 070	
Passengers (Centre Seats) or					
Baggage Zone 2 (360 lb Max).	360.00	118.5	42 660.00	42 660	
Passengers (Rear Seats) or					
Baggage Zone 3 (400 lb Max).	240.00	157.6	37 824.00	37 824	
Baggage Zone 1 (100 lb Max).	100.00	22.5	2250.00	2250	
Baggage Zone 4 (100 lb Max).	77.00	178.7	1375.99	13 759.9	
Zero Fuel Mass (4470 lb Max)	**4327.00**	**91.8**	**397 264.99**	**409 648.9**	**94.7**
Fuel (123 gal Max)	444.00	93.6	41 558.40	41 558.4	
Ramp Mass (4773 lb Max)	**4771.00**	**92.0**	**438 823.39**	**451 207.3**	**94.6**
Start Fuel	-22	93.6	-2059.20	-2059.2	
Take-Off Mass (4750 lb Max).	**4749.00**	**92.0**	**436 764.19**	**449 148.1**	**94.6**
Minus trip fuel	-400.00	93.6	-37 440.00	-37 440	
Landing Mass (4513 lb Max).	**4349.00**	**91.8**	**399 324.19**	**411 708.1**	**94.7**

Chapter 10

Medium-Range Jet Transport (MRJT 1)

INTRODUCTION

This chapter introduces the MRJT 1 as per CAP 696 JAR FCL Examinations Loading Manual. The aircraft is a medium-range twin jet certified under FAA/JAR-25 in performance class A.

The MRJT 1 data sheets in the CAP 696 (pages 19 to 31) show in an abbreviated format the type of loading manual documentation that pilots are likely to encounter on a large transport aircraft.

Read these notes in conjunction with the CAP 696 in order to become familiar with the presentation and use of the data. The notes use extracts from the CAP 696. Where these are used, the Figure Number or Table Number along with the page number are given.

CONTENTS (PAGE 19)

The manual is split into seven sections:

1. Aircraft Description
2. Aircraft Data Contents
3. Mass and Balance Limitations
4. Fuel Data
5. Passengers and Personnel Data
6. Cargo Data
7. Mass and Balance Calculations

LOCATIONS DIAGRAM (FIGURE 4.1, PAGE 20)

This diagram shows the aircraft viewed from the right wing tip. The datum's location in the nose of the aircraft is found by measuring 540 inches forward from the front spar. On the diagram, this is marked FS below the fuselage and 540 above the fuselage. See fig. 2.1, datum point at the bottom of the page.

TABLE TO CONVERT BODY STATIONS TO BALANCE ARM (FIGURE 4.2, PAGE 20)

Body Station	Conversion	Balance Arm-in
500A	348 + 22 in	370

To find a Balance Arm (BA), enter the table for a given Body Station (BS), for example, BS 500A, and read across to the middle column. This column gives the conversion factor. In this case, a constant of 348 and a 22-inch pitch increase for this row, giving a BA of 370 in.

LANDING GEAR RETRACTION (PAGE 21)

Operation of the landing gear has negligible effect on the CG.

EFFECT OF FLAP RETRACTION (FIGURE 4.3, PAGE 21)

From	To	Moment Change (kg inches x 1000)
5	0	-11
15	0	-14
30	0	-15
40	0	-16

In figure 4.3, the effect of raising the flaps is given as an index number. This is negative when the flaps are retracted and positive when they are lowered. For example, retracting the flaps from 30° to 15° results in a -1000 kg in moment (15 – 14) x 1000.

TAKE-OFF HORIZONTAL STABILISER TRIM SETTING (FIGURE 4.4, PAGE 21)

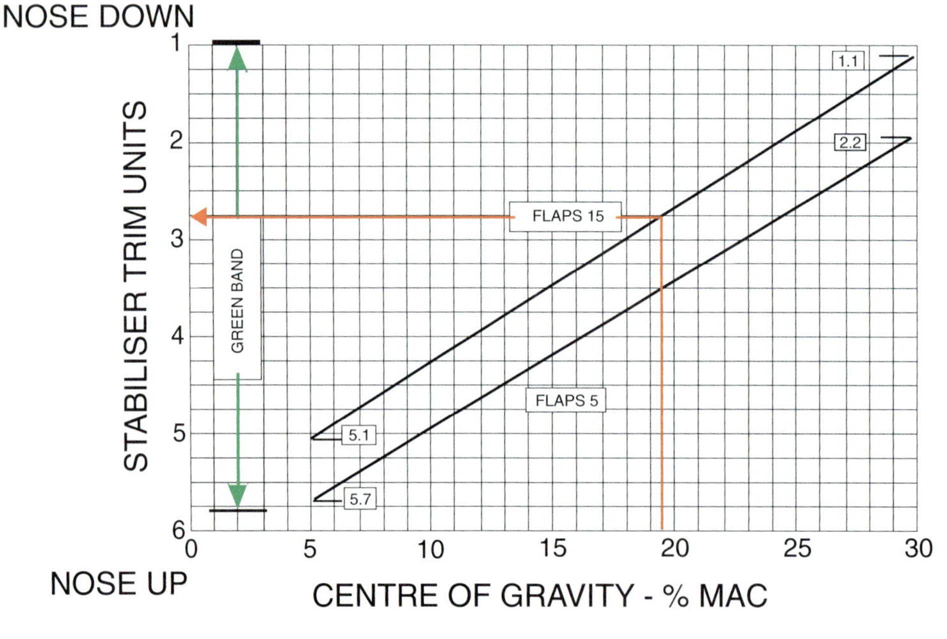

Figure 4.4 is a graph showing horizontal stabiliser trim settings for 5° and 15° of flaps against CG positions given in % MAC. This graph is used to find the stabiliser trim required depending on the CG position and flap setting. For example, use the graph, shown in figure 4.4 of the data sheet, to determine the stabiliser setting with 15° flaps and CG at 19.5% MAC.

1. Enter graph at 19.5 % MAC and draw a vertical line to intersect the 15° of flaps line.

2. Then draw a horizontal line again parallel to the grid to the left axis.

3. Read off the side scale to obtain the correct stabiliser trim units, in this case 2.8.

CONVERSION OF BA TO OR FROM % MAC (PAGE 21)

The Mean Aerodynamic Chord for the aircraft is given as 134.5 in with its leading edge 625.6 in aft of datum.

The conversion of a linear CG position uses the standard formula:

$$\frac{A - B}{C} \times 100 = \% \text{ MAC}$$

For the MRJT 1, insert the values of 625.6 for B and 134.5 for C, as shown below:

$$\frac{A - 625.6}{134.5} \times 100 = \% \text{ MAC}$$

Example

The aeroplane is weighed in order to determine the BEM and related CG. The following readings are obtained (in kiloNewtons):

Location	**BA (inches)**	**Force (kN)**
Nose Wheel	158	29.95
L Main wheel	698	152.45
R Main wheel	698	153.10

Determine the BEM in kg and the CG as a % MAC.
Assuming the gravitational constant G = 10 m/sec^2

1. Determine total force and total moment as shown in the table below.

Location	**BA inches**	**Force kN**	**Moment kN/in**
Nose	158	29.95	4732.1
Left Main	698	152.45	106 410.1
Right Main	698	153.1	106 863.8
	Totals	**335.5**	**218 006.0**

BEM = (335.5 X 1000) ÷ 10 m/s^2 = 33 550 kg

2. Finding the CG as % MAC
CG = Total Moment ÷ Total Force
= 218 006 kN/in ÷ 335.5 kN
= 649.8 inches

$$\frac{649.8 - 625.6}{134.5} \times 100 = 17.99 \% \text{ MAC}$$

= 18% MAC

MASS AND BALANCE LIMITATIONS (PAGE 21)

The following are given:

Maximum structural taxi mass	63 060 kg
Maximum structural take-off mass	62 800 kg
Maximum structural landing mass	54 900 kg
Maximum structural zero fuel mass	51 300 kg

CENTRE OF GRAVITY LIMITS (PAGE 22)

The CG limits are shown in a graphical form in figure 4.11, CG Envelope, page 27.

FUEL (PAGE 22)

Fuel loads and limits for the aircraft are given in paragraph 4.

Figure 4.5 shows fuel tank locations and capacities, giving balance arms (inches) and quantities in both US gallons and kilograms. The SG is 0.8 for the figures given.

Note the caution below the table about the centre tank, and that the mass of fuel is given as 3.03 kg per US gal.

During flight, the weight in the fuselage combined with lift produced by the wings (airload) causes them to bend upward. To counter this, it is standard practice to use the fuel from the centre tanks first, then the wing tanks, working progressively outward. This reduces the fuselage mass, which reduces the amount of lift required and balances this lift with the weight of fuel in the wing tanks, reducing the bending effect in the wing structure.

Figure 4.6 shows the same information as figure 4.5 for unusable fuel. The order of the columns is changed.

A tank location diagram is shown at the bottom of the page.

PASSENGERS (PAX) AND PERSONNEL (PAGE 23)

Details on passenger (pax) and personnel are given on page 23. These are:

Maximum Passenger Load	141
Club/Business	33
Economy	108

Paragraph 5.2 and the associated diagrams, figure 4.7 and figure 4.8, detail the passenger distribution in the cabin. Note the comment about seating for low passenger loads.

In figure 4.8, the table details the max capacity and balance arm, and the **centroid** of each zone. The BA given is the arm length for the mid-zone position.

PASSENGER MASS (PAGE 23)

Paragraph 5.3 of the data sheet states to assume the passenger mass as 84 kg unless otherwise stated which includes 6 kg hand baggage. As there is no mention of passenger age or gender in this section of the data sheet, assume that every passenger weighs 84 kg.

PASSENGER BAGGAGE (PAGE 23)

The passengers baggage mass is given as 13 kg.

PERSONNEL (PAGE 23)

Standard Crewing	Number	BA	Standard Mass kg
Flight Deck	2	78.0	90
Cabin Staff - Forward	2	162.0	90
Cabin Staff - Aft	1	1107.0	90

All crewmembers regardless of gender are given a standard mass of 90 kg.

CARGO (FIGURE 4.9, PAGE 24)

Page 24 details the aircraft's front and rear cargo compartment limitations. These take the form of two tables. In figure 4.9, which is titled Cargo Compartment Limitations, the complete space is normally referred to as a hold and is subdivided into compartmental areas.

Forward Cargo Compartment

BA – IN 228 286 343 500

MAXIMUMCOMPARTMENT RUNNING LOAD IN KG PER INCH	13.15	8.47	13.12
MAXIMUM DISTRIBUTION LOAD INTENSITY KG PER SQUARE FOOT	68		
MAXIMUM COMPARTMENT LOAD KG	762	483	2059
COMPARTMENT CENTROID BA - IN	257	314.5	421.5
MAXIMUM TOTAL LOAD KG	3305		
FWD HOLD CENTROID BA - IN	367.9		
FWD HOLD VOLUME CUBIC FEET	607		

The forward cargo hold is divided into three compartments, each with a different running load limit but with the same static load limit (intensity).

Maximum Compartment Load

The length of the compartment multiplied by the running load determines the maximum compartment load. For example, the fwd compartment of the fwd hold is 58 inches long (286 - 228) and the running load is 13.15 kg/in. 58 in X 13.15 kg/inch = 762.7 kg, which is given as 762 kg in the table.

Centroids

Referring to fig. 4.9 of the data sheets, any mass acting in the fwd hold can be said to act from the centroid arm of 367.9 in. However, each compartment of the hold has its own centroid. In the case of the fwd compartment, this is a balance arm of 257 in aft of the datum. The question will guide the candidate as to which arm is to be used.

Note: If referred to the data sheets by any question, be sure to use the numbers given in the appropriate data sheet.

LOADING MANIFEST (FIGURE 4.10 & 4.11, PAGE 25)

Paragraph 7.1 details how to calculate the mass and balance for the aircraft using the loading manifest figure 4.10 on page 26 and the CG envelope figure 4.11 on page 27. Read the list as this ensures that the aircraft's mass and balance is checked against the limits.

The fuel arm is taken from the table on page 22.

For the following worked example, a loading manifest has been completed to illustrate its use. The data for the structural limits have been taken from the data sheet. The passenger, cargo, and fuel loads have been made up and are given in the table. Fuel consumption is given below:

Figure 4.10 Loading Manifest – MRJT 1

Max Permissible Aeroplane Mass Values:

TAXI MASS - 63 060 kg ZERO FUEL MASS - 51 300 kg

TAKE-OFF MASS - 62 800 kg LANDING MASS - 54 900 kg

ITEM	MASS (kg)	BA IN	MOMENT kg in/1000	CG %MAC
1. DOM	34 500	649.00	22 390.5	
2. PAX Zone A	840.00	284.00	238.60	-
3. PAX Zone B	1512	386.00	583.60	-
4. PAX Zone C	2016	505.00	1018.1	-
5. PAX Zone D	2016	641.00	1292.3	-
6. PAX Zone E	2016	777.00	1566.4	-
7. PAX Zone F	1512	896.00	1354.8	-
8. PAX Zone G	1092	998.00	1089.8	-
9. CARGO HOLD 1	650.00	367.9	239.10	-
10. CARGO HOLD 4	2120	884.5	1875.1	-
11. ADDITIONAL ITEMS	NIL	N/A	NIL	-
ZERO FUEL MASS	48 274	655.60	**31 648.3**	**22.30**
12. FUEL TANKS 1 & 2	9084	650.70	5911	-
13. CENTRE TANK	4916	600.40	2951.6	-
TAXI MASS	62 274	650.50	**40 510.9**	**18.50**
LESS TAXI FUEL	-260.00	600.40	-156.10	-
TAKE-OFF MASS	62 014	650.70	**40 354.8**	18.70
LESS FLIGHT FUEL	4844 4656 Total 9500	650.7 600.4	-3152 -2795.5	-
EST. LANDING MASS	52 514	655.20	**34 407.3**	22.53

Centre Tank has 4916 kg at start-up and is used first. The taxi allowance of 260 kg is subtracted, leaving 4656 kg in the tank for take-off. For take-off and in the initial stages of flight, the trip fuel is drawn from the centre tank first. When this tank is drained, the remaining trip fuel is drawn from the wing tanks. (9500 – 4656) = 4844 kg from wing tanks. Having completed the manifest and checked the ZFM, TOM, and LM against the limitations, these masses are plotted against the CG as a % of MAC on the CG envelope.

Figure 4.11 CG Envelope – MRJT 1

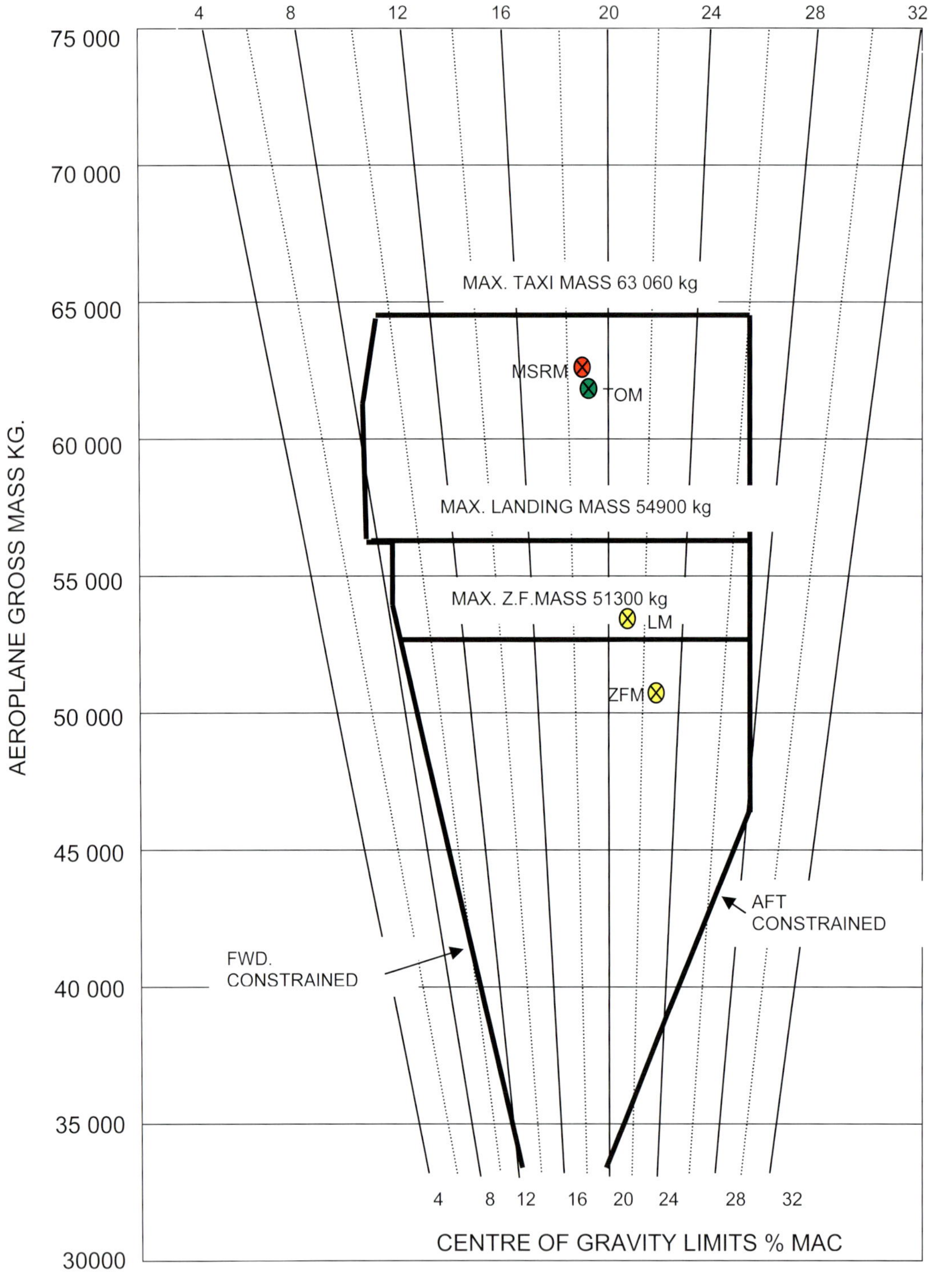

When all masses are found to be in the stated limits and the CG points have been plotted on the CG envelope (fig. 4.11), the question "Is the aircraft in limits?" can be answered.

In fig. 4.11, the CG envelope shows both forward and aft limits in terms of % MAC at Gross Masses from 30 000 kg to 63 060 kg, the limits for MLM and MZFM. Note the change in shape of the envelope and the sharp reduction in the aft CG limit as gross mass drops below 44 250 kg.

THE LOAD AND TRIM SHEET (FIGURE 4.12 & 4.13, PAGES 28 ONWARD)

Load and trim sheets are a method of calculating an aircraft's mass and balance. Operators of larger transport aeroplanes use them to:

- Speed up the process of Mass and Balance calculations
- Provide the flight crew with the essential information in an easy to use format
- Provide the necessary documentation as required by the Authorities

The load sheet required by JAR-OPS, Subpart J includes the following mandatory information:

Aeroplane registration and type
Flight number
PIC
The identity of the loader
The DOM and CG position
Mass of take-off fuel and trip fuel
Mass of consumables other than fuel
Traffic load, passengers, baggage, and freight
TOM, LM, and ZFM
Load distribution
Aeroplane CG positions
Limiting mass and CG values

Typical of such documentation is the Load and Trim Sheet, an example of which appears in the CAP 696 as figure 4.14.

The procedure for using the Load and Trim Sheet in CAP 696 is given in paragraph 7.2, page 28, with a worked example on page 29 as figure 4.12.

The example Load and Trim Sheet, figure 4.12, is broken down into two areas:

- Part A is the loading summary and is used to derive all the weights from DOM to LM.
- Part B is the trim portion on which movements of CG may be derived for each mass from DOM to LM and includes elements for each portion of the load.

Part A is divided into 3 sections.

Section 1 is used to establish the limiting TOM, maximum allowable traffic load, and under load before any last minute changes.

Section 2 gives the distribution of the Traffic Load.

TR	Transit
B	Baggage
C	Cargo
M	Mail
Pax	Passengers
Pax F	First Class passengers
Pax C	Club/Business Class passengers
Pax Y	Economy Class passengers

Section 3 is a summary of the loading and is a crosscheck that limiting values are not exceeded. The abbreviations used include:

DOI	Dry Operating Index
MLDGM	Maximum Landing Gross Mass

Note: For the following examples, using the load and trim sheet in the CAP 696 and these notes, the baggage is given a standard mass of 14 kg. For the loading manifest data, 13 kg is used.

FUEL INDEX CORRECTION (FIGURE 4.13, PAGE 30)

Figure 4.13 is a correction table tabulating index movement against fuel mass.

Note: The comments below the table are important when attempting any questions.

Illustrations of the load and trim sheets and a step-by-step approach on how to use them are given in the following pages. At first sight, it may appear to be a very complicated presentation. With practice, they do become “user-friendly”.

Figure 4.14 Load and Trim Sheet (Blank)

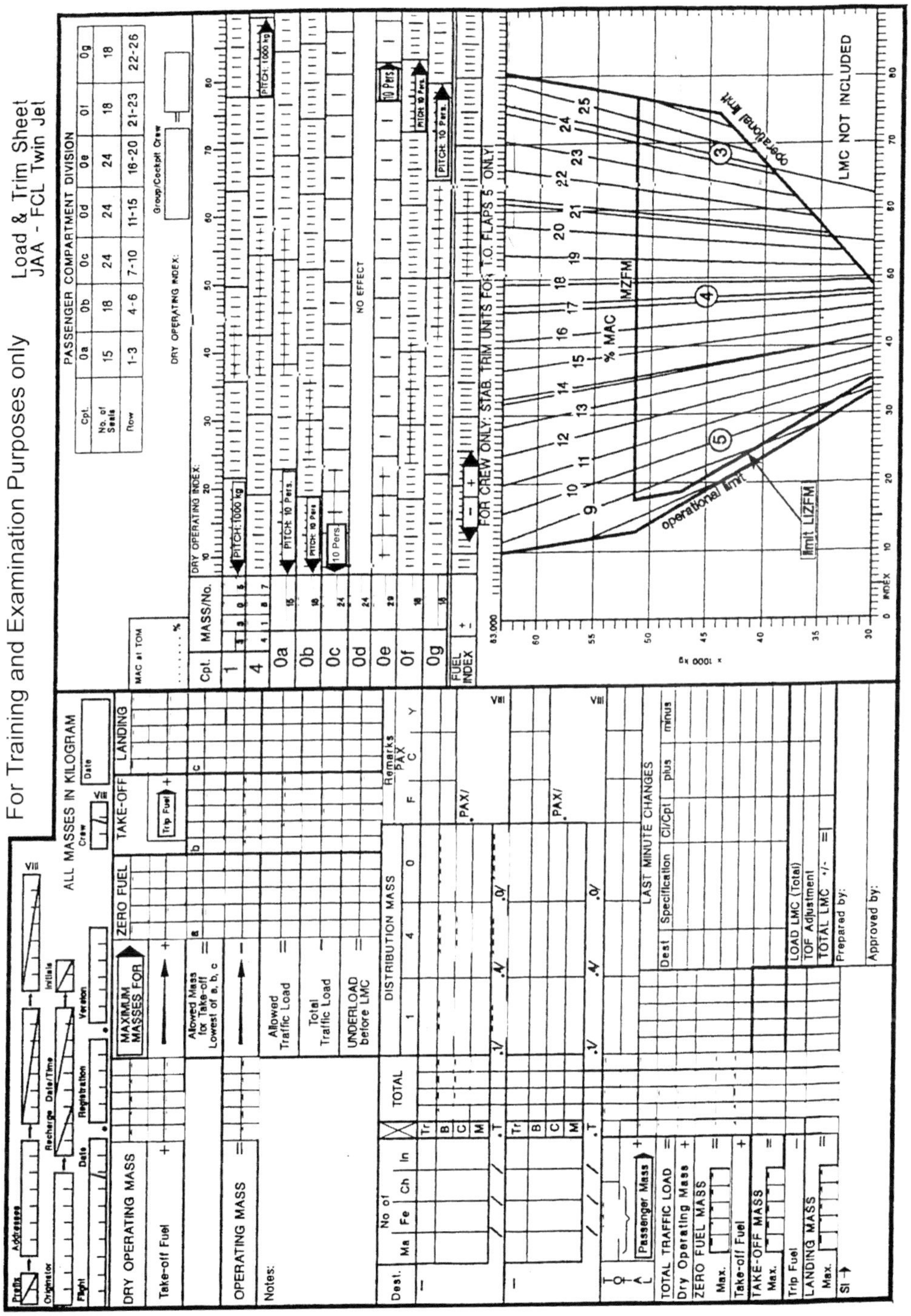

THE LOAD SHEET

To complete the load sheet, follow the instructions in conjunction with the enlarged excerpts of figure 4.12, CAP 696. This has been divided into the three sections for clarity. Values used are those of figure 4.12.

SECTION 1

This section of the load sheet is divided into 4 columns. In the second column, there are two arrows pointing to the right. These show where values from the first column are repeated in other columns. In this section, there are + and – signs printed, and the calculations run vertically downward.

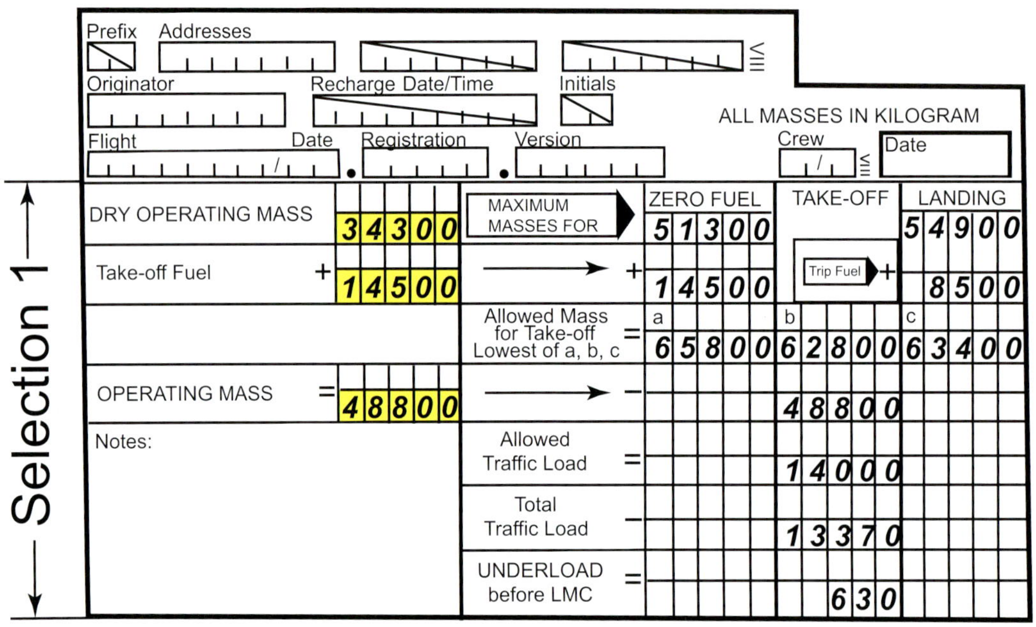

1. Enter the DOM 34 300 kg.

2. Below the DOM, enter the take-off fuel, 14 500 kg.

3. Add the DOM and the TOF together to find the OM, 48 800 kg. Enter this value against the operating mass.

2nd column and 3rd column

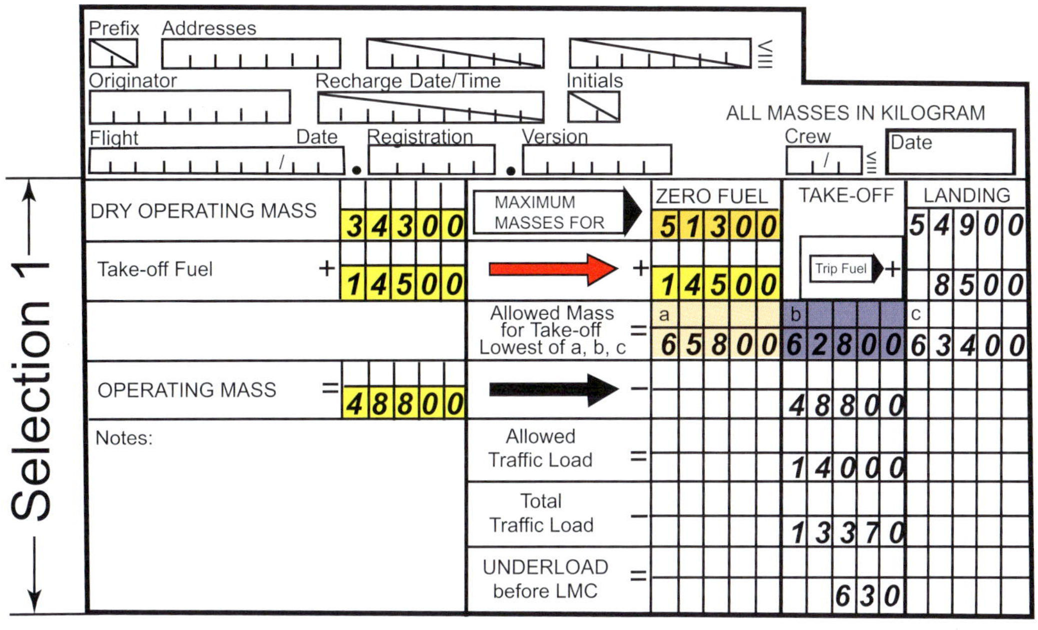

4. Enter the MZFM 51 300 kg and enter the TOF 14 500 kg below it.

5. Add the MZFM and TOF together, 65 800 kg. Enter this in the next line down below the (a) against the heading Allowed Mass for Take-Off. The lowest of a, b, and c is the Allowed Mass for Take-Off.

6. Below the (b) in the row against "Allowed Mass for Take-Off/Lowest of a, b, c" heading, enter the lowest of MTOM, PLTOM, or RTOM. This has been given as 62 800 kg.

4th column

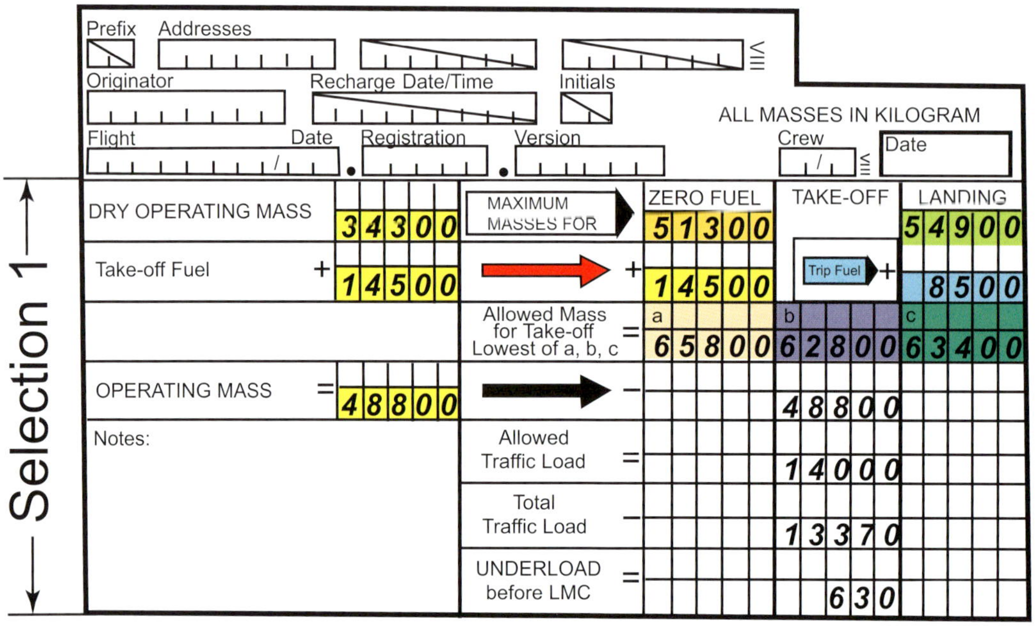

7. Under the heading Landings, the lowest value of MLM, PLLM, or RLM is entered. This has been given as 54 900 kg. Below this is entered the trip fuel mass given as 8500 kg.

8. Add the landing mass to the trip fuel mass 63 400 kg and enter this value below the (c) in the row against Allowed Mass for Take-Off/Lowest of a, b, and c.

TOM and Traffic Load

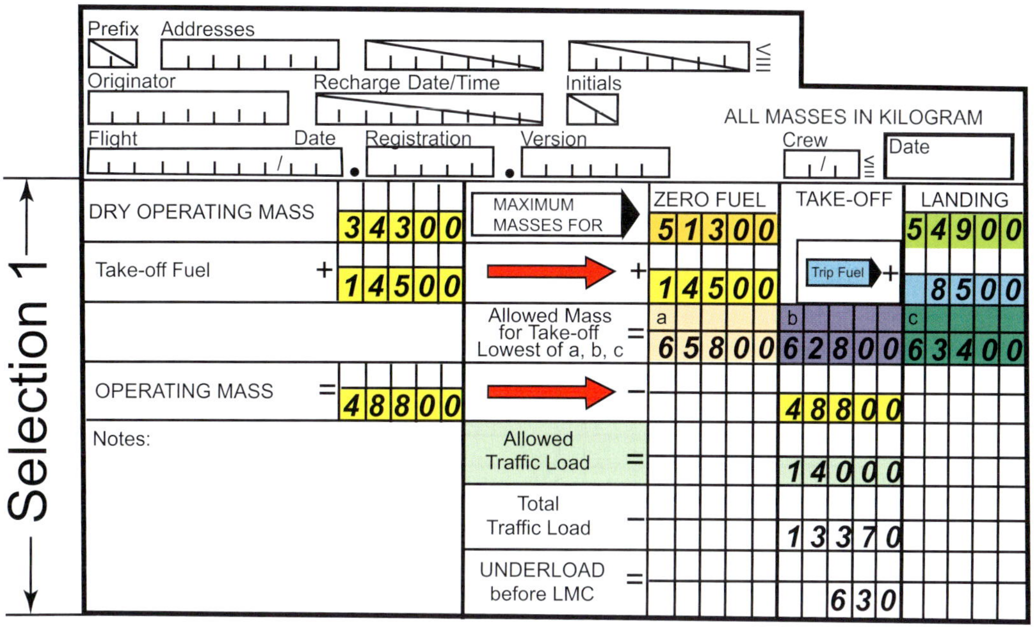

9. Select the lowest value in the row Allowed Mass for Take-Off. In this case, it is (b) 62 800 kg and is referred to as MATOM (Maximum Allowed Take-Off Mass).

10. Use the following calculations to find the allowed traffic load, total traffic load, and under load. These are carried out in the MATOM column, in this case b.

11. The OM 48 800 kg is carried across and entered in the column and subtracted from the MATOM 62 800 kg. The difference, 14 000 kg, is entered against the Allowed Traffic load. This is maximum traffic mass for an aircraft of this DOM and TOF, taking off from this departure airport and landing at the destination airport.

SECTION 2

The distribution of the traffic load is calculated in this section. Note that the area is divided into three main columns.

Items 12 to 14

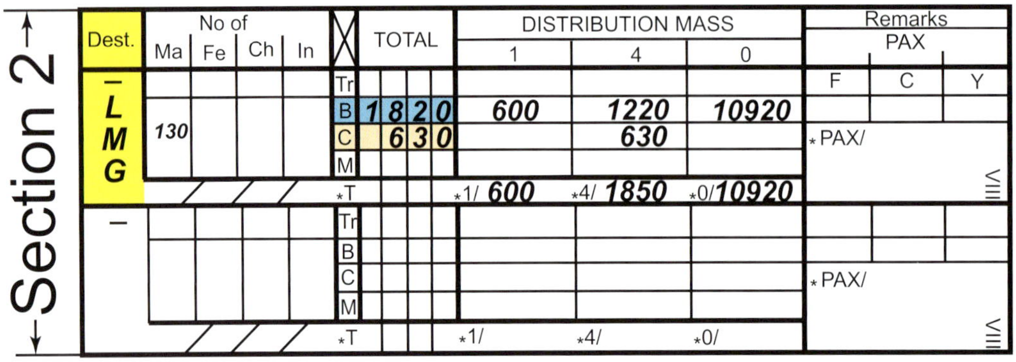

12. In the first column, the passenger details are given in seven sub-columns. The first is labelled Dest for destination. The three-letter code for the destination airport is entered here, in this case LMG.

13. Across from Dest are four sub columns headed Ma, Fe, Ch, and In for Males, Females, Children, and Infants. The number of each is entered in the appropriate sub-column. As in this case, the aircraft is using the all adult masses. The total number of passengers (130) is entered under Ma.

14. The sixth sub-column gives the code letters Tr (Transit), B (Baggage), C (Cargo), and M (Mail) as per page 28 of the CAP 696. Against the appropriate code letter, the mass is entered in the seventh sub-column. In this case, 1820 kg for baggage and 630 kg for cargo.

Items 15 to 19

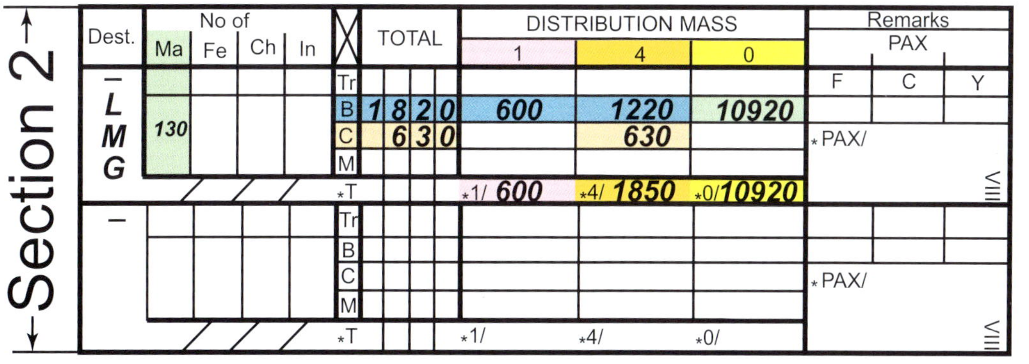

15. The second column is headed distribution of mass and divided into 3 sub-columns headed 1, 4, and 0. Sub-column 1 is the forward cargo hold, sub-column 4 is the aft cargo hold, and sub-column 0 is the passenger cabin.

16. In this example, the baggage 1820 kg has been divided between the two holds with 600 kg placed in the fwd hold and 1220 kg in the aft hold. The cargo of 630 kg is also placed in the rear hold. The passenger mass for the cabin is entered as 10 920 kg.

    130 pax times the average mass of 84 kg = 10 920 kg

17. Below code letter M in the first column is *T. This is the total line for the rows above: *1/ 600 kg, * 4/ 1850 kg and * 0/ 10 920 kg.

18. If any small mass other than pax is to be carried in the main cabin, it would be entered in the 0 sub-column. The pax load mass can be entered by gender and age to assist in checking the calculation.

19. The third column headed remarks/pax is divided into three sub-columns: F (First Class), C (Club/Business), and Y (Economy). These are the class codes as per page 28 of the CAP 696.

SECTION 3

This section compares the mass totals with the limits to ensure that they are not exceeded. The section is divided into two main areas:

Left side As loaded

Right side Details any last minute changes and has the signature blocks

Items 20 to 27

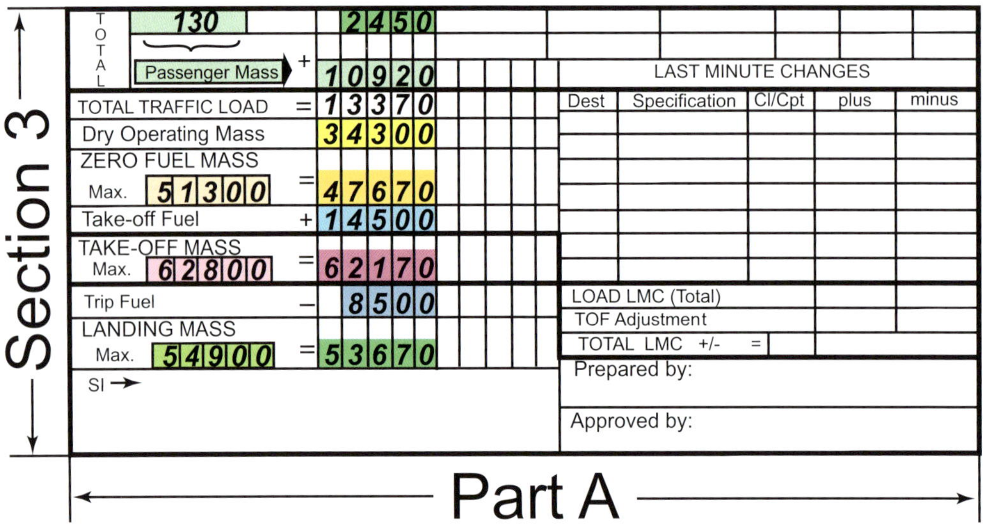

20. In the first column, the total number of passengers is entered (130). To the right in the second column is a total mass of the baggage and cargo, etc., from section 2. In this case, it is 2450 kg.

21. Below this, the passenger mass (10 920 kg) from section 2 is entered. These totals are added together to find the traffic load (2450 kg + 10 920 kg = 13 370 kg). This is entered on the next line down.

22. The aircraft's DOM (34 300 kg) is entered on the next line down.

23. The MZFM (51 300 kg) is entered in the first column. Then, the Total Traffic Load and the DOM are added to find the aircraft's ZFM (13 370 kg + 34 300 kg = 47 670 kg), which is entered in the second column.

24. The Take-Off fuel mass (14 500 kg) is entered below the ZFM in the second column and added to the ZFM (14 500 + 47 670 = 62 170 kg) to find the TOM 62 170 kg.

25. The MATOM (62 800 kg) is entered in the first column.

26. The trip fuel (8500 kg) is entered in the second column and subtracted from the TOM to find the LM (62 170 kg – 8500 kg = 53 670 kg). This is entered in the second column.

27. The Maximum Allowed Landing Mass (54 900 kg) is entered in the first column.

RETURN TO SECTION 1

Items 28 and 29

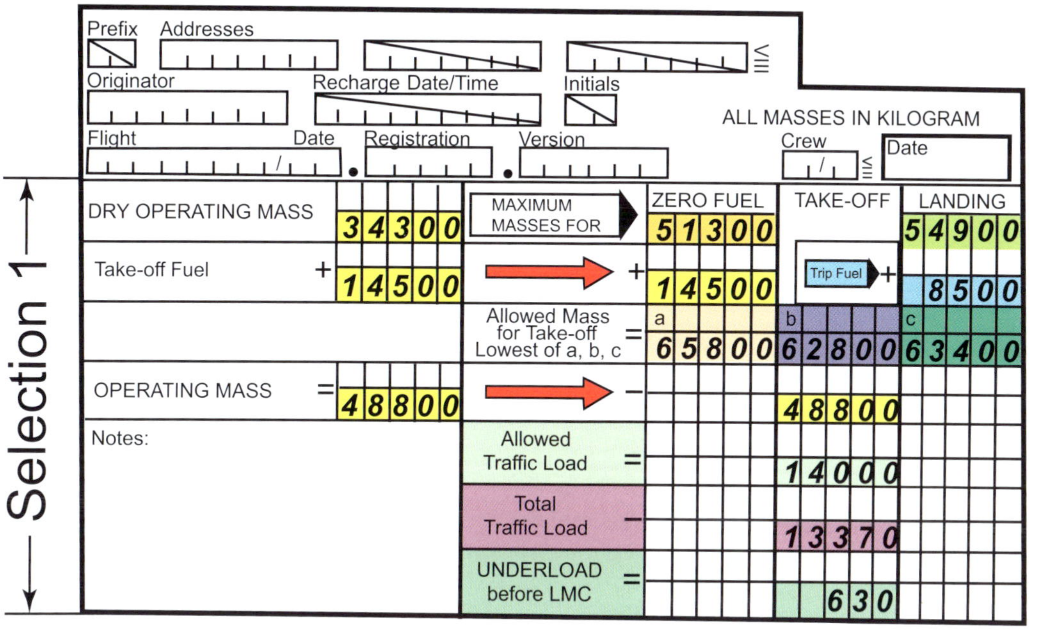

28. The actual traffic load (13 370 kg) is entered in the next row down against the Total Traffic Load.

29. The Total Traffic Load value (13 370 kg) is subtracted from the Allowed Traffic Load Value (14 000 kg). The 630 kg difference is the amount of under load. This mass could change if there are any Last Minute Changes such as passengers or freight being loaded or off loaded.

All the limitations entered in the 1st column of Section 3 are those that have been worked out in Section 1. The masses for the Total Traffic Load are worked out from those masses entered in Section 2, and the Take-Off Fuel and Trip Fuel are those given in Section 1.

This allows the compiler and Commander to cross check the data.

LAST-MINUTE CHANGES

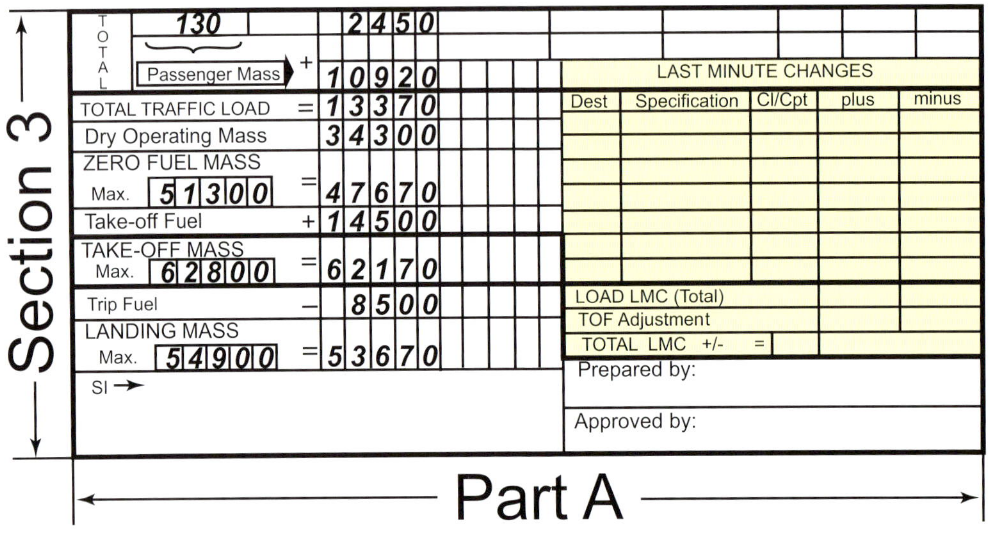

Enter the total mass of LMC (LMC payload + change to TOF). Then check to ensure that the figure does not exceed the allowable extra mass. Last-minute changes must be brought to the attention of the Commander.

TRIM SHEET

The trim sheet, figure 4.12, consists of two main areas:

The top half is effectively a series of horizontal scales above a CG envelope. The trim sheet uses moment index numbers to keep the figures used within manageable limits.

To demonstrate the method of using the Trim Sheet, the CAP 696 example, figure 4.12, is divided into the two component parts. The top section of the Trim Sheet is shown below.

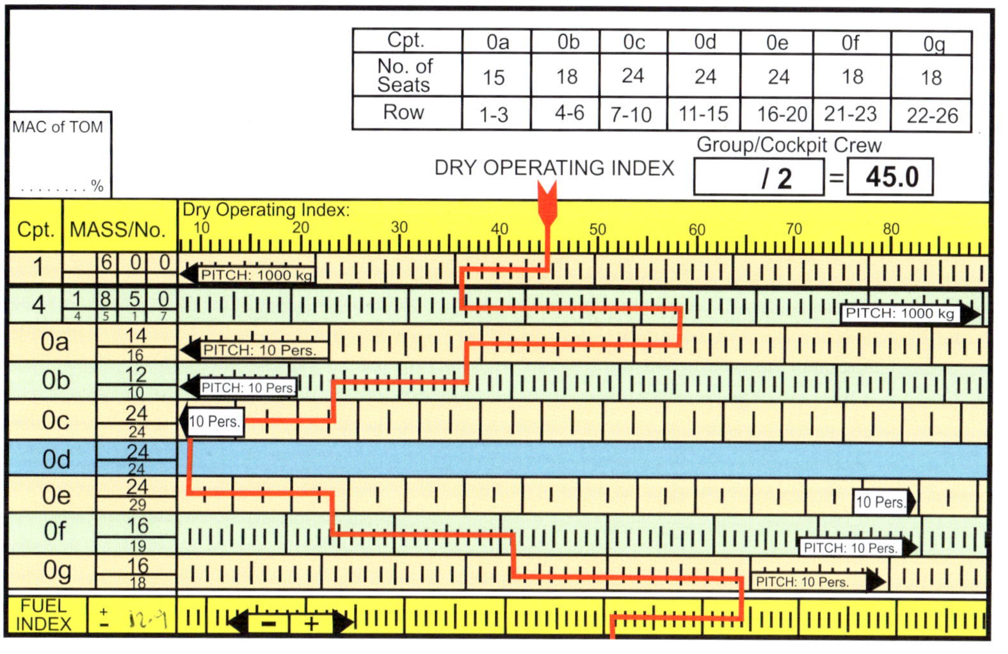

Before using the Trim Sheet, a few details need to be highlighted.

The table at the top right of the page shows the breakdown of the passenger cabin into areas Oa, Ob, etc. (Refer to figure 4.7 of the CAP 696 for a pictorial view of the aircraft.) Under each area is the maximum number of seats and the number of rows. For Oa, there are 15 seats in rows 1 to 3.

Below the passenger compartment table is a box titled Group/Cockpit Crew. The group code or number of flight crew is entered into this box, and the DOI is entered into the second box.

Note: "Group" refers to the operator's configurations and is unlikely to be used in examination. However, a DOI value will be given if required by the question.

To the left is a box titled MAC at TOM. This % is found when the trim sheet is completed.

Below the MAC at TOM box are two columns titled Cpt and MASS/No respectively. Cpt is an abbreviation for compartment.

In the first column, Cpt, each row indicates a compartment centroid arm; 1 is the fwd hold, 4 is the aft hold and then progressing rearward through the cabin.

In the second column, each row is sub-divided horizontally. The lower portion has the limiting mass or limiting number for the compartment, as shown below for compartment Oa.

Oa	
	15

To the right of these columns are the main scales for Cpt 1 to Og. Each scale varies from line to line. This reflects the effect that a given mass has in each compartment.

Note: Compartment Od has no effect on the CG (refer to paragraph 5.2 on page 23 CAP 696).

In each row, there is an arrow denoting the direction in which the CG moves as mass is added. The pitch scale is printed in the body of the arrow. For Cpts 1and 2, it is in kilograms, for Cpts Oa to Og, it is the number of passengers.

Example

1. Enter all the known details from the load sheet for Cpt 1 + 2. 600 and 1850

2. Enter the number of passengers per compartment for Oa to Og given as:

Oa	14
Ob	12
Oc	24
Od	24
Oe	24
Of	16
Og	16

3. Check to ensure that mass and pax limits are not exceeded, and that the total number of pax agrees with the number given in section 2 of the load sheet for the worked example 130.

4. Enter the number of cockpit crew in the box marked Group/Cockpit Crew. For the example, the crew is given as 2.

5. In the box to the right, enter the Dry Operating Index DOI. In this case, the Index is given as 45.0 as per the data sheet page 28.

6. In the top scale titled Dry Operating Index, find the DOI and mark it. In this case, it is 45.0.

7. Drop a vertical line from the mark 45.0 DOI into the centre of the horizontal scale below Cpt 1.

8. In this scale, the arrow is pointing left and the pitch is given as 1000 kg per large division. Therefore, each small division is equal to 100 kg. The cargo mass of 600 kg is equal to a horizontal movement to the left of six small divisions.

9. Where the vertical line dropped from a previous scale does not directly match the scale line. The compiler must measure the exact distance to be moved from the point of entry. As can be seen from the line in figure 4.G, on completing the measurement, a vertical line is dropped into the centre of the next row.

10. The operator continues the sequence and where the mass or number of pax differ from the given scale, the operator has to interpolate to find the exact distance to move horizontally.

11. On completing the compartments, the aircraft's ZFM CG can be found by dropping a vertical through the fuel index row into the CG envelope, as shown below.

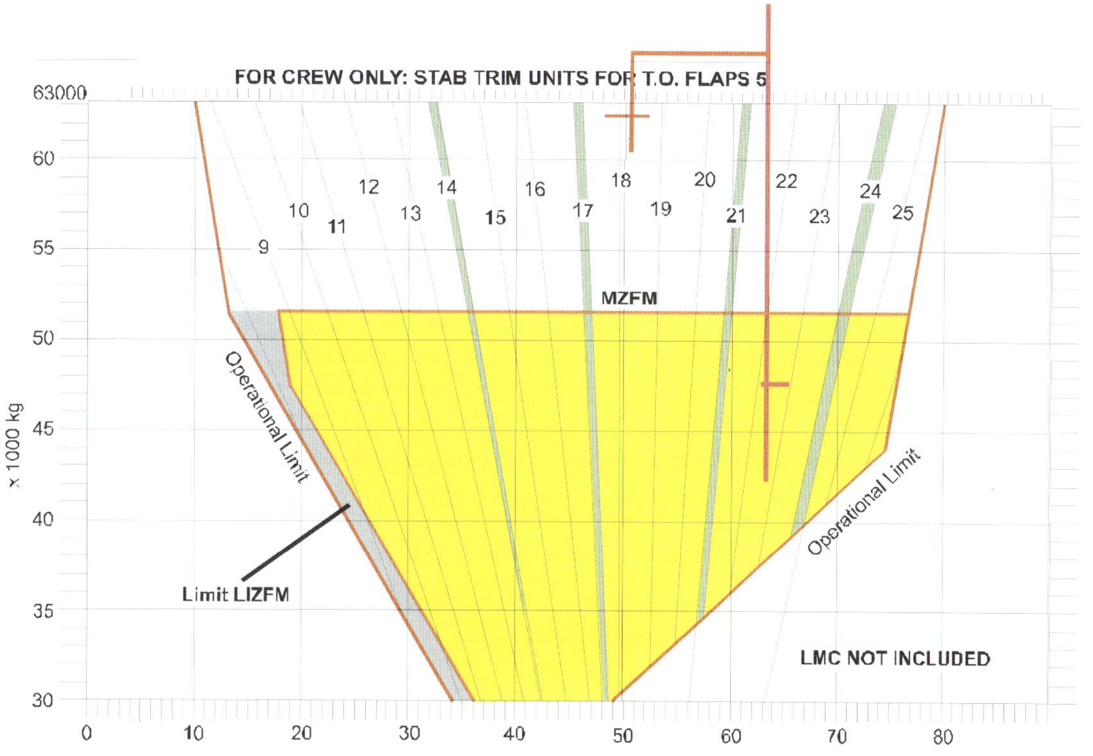

12. Take the ZFM from the load sheet, in this case 47 670 kg, and find this point on the vertical scale at the side of the envelope. Draw a horizontal line from this point through the vertical index line. The ZFM CG is located where these lines intersect and can be read off as a % MAC from the envelope's scale.

13. Check that this intersection is within the LIZFM limits. (Load Index ZFM) and the MZFM limit.

Note: For this aircraft, the Fwd limit for the ZFM is less than the operational limits (front and rear limit of the safe range).

14. To add the fuel and account for its effect, the compiler has to refer to the Fuel Index Correction Table, figure 4.13 on page 30 of the CAP 696. In the worked example, the take-off fuel load is 14 500 kg. As this cannot be found directly, the compiler reads the next higher mass 14 580 kg giving an index of 12.9.

The fuel index row has a double arrow to the left for negative index units and to the right for positive index units. The scale has a pitch of 10 units per large division.

15. To find the take-off index, the fuel index units are added or subtracted from the ZFM index. In the case of the worked example, the units are negative. Therefore, a horizontal line is drawn for 12.9 divisions to the left. A vertical line is dropped into the CG envelope where a horizontal line for the TOM is drawn to intersect the vertical. This shows the TOM CG.

16. Check that the TOM CG is within the operational limits.

If a line is drawn connecting the centres of gravity for the TOM and ZFM conditions, the CG should move progressively down this line as the usable fuel is consumed.

To find the Landing Mass CG, take the trip fuel mass from the Take-Off fuel mass to find the fuel remaining in the aircraft's tanks on landing.

In this case, 6000 kg (14 500 kg - 8500 kg = 6000 kg) and convert it into an index unit value using the table, figure 4.13 on page 30.

For 6000 kg, this is –6. This is plotted in the fuel index scale from the ZFM vertical line. Then a vertical line is dropped into the CG envelope. The Landing Mass CG is located where the vertical line bisects a horizontal line plotted for the Landing Mass 54 900 kg. This should also bisect the line joining the TOM and ZFM.

Note: The CG locations in the envelope do not include any LMCs.

Other uses for the trim sheet are:

- Finding the index number for an aircraft at a given CG condition
- Finding the exact CG as a % MAC
- Finding the OM CG location
- Adjusting the CG's location

FINDING THE INDEX NUMBER FOR AN AIRCRAFT AT A GIVEN CG CONDITION

To find the index number for any CG location, drop a vertical line from the CG's position in the envelope to the index scale below it and read the value.

Where the CG falls between two given values of MAC, in the case of the worked example between 18 and 19% MAC for the TOM, measure the distance between these two lines level with the CG's location and the distance between the CG and line before it (18% MAC in this case). Work the fraction into a decimal (e.g. 2/6 mm = 33%). The CG is, therefore, located at 18.33% MAC (obviously the larger the actual size of the envelope, the more accurate the reading). This level of accuracy is not likely to be required.

FINDING THE OM CG LOCATION

If the operating mass CG is to be found, enter the DOI at the DOI scale. Then drop a vertical line down into the fuel index scale. Add or subtract the fuel index value to move horizontally. Then drop a vertical line into the envelope and bisect this with a horizontal for the OM.

ADJUSTING THE CG'S LOCATION

If there is a need to relocate the CG from its current position to a new location, the trim sheet can be used to work out the amount of cargo, baggage, or passengers that is required to be moved or off loaded, etc. The great advantage of the trim sheet is that the effect can be seen. It is easier to see the workings for removal or addition before looking at load shifting.

The method used is:

Draw a vertical from the current CG position to the bottom of the DOI scale. Draw another vertical line from the intended CG location again to the bottom of the DOI scale.

Note: Do not fall into the trap of aligning the rule with the % MAC lines as these diverge. Use the grid lines.

Read off the difference between the lines at each scale. Note the direction in which the arrow points. This indicates the amount that must either be removed from the aircraft or added to the aircraft for that compartment to alter the CG's location.

Use the worked example to relocate the CG from its current position to a new location of 18% MAC. After drawing the two vertical lines, it would be necessary to:

- Add 100 kg to the fwd hold.
- Add a 100 kg to the aft hold.
- Add 1 Pax to Cpt Oa.
- Add 1 Pax to Cpt Ob.
- Add 1 Pax to Cpt Oc.
- Alterations in Cpt Od have no effect.
- Remove 1 Pax from Cpt Oe.
- Remove 1.5 Pax from Cpt Of.
- Remove 1 Pax from Cpt Og.

Any single alteration from the above list causes the CG to relocate to 18% MAC.

Where removing or adding is not practicable or commercially desirable, the load would have to be relocated. Determine how much load is to be shifted (and to which location it is to be shifted) in order to effect a change of trim.

Note: Additions and subtractions alter the Gross Mass for all the load conditions.

Method:

- Note the direction in which the CG is to move. If this is fwd, the mass to be relocated is taken from the rear hold and placed in the fwd hold and vice versa.
- As before, draw verticals from the two points in the envelope up through hold 4 and hold 1 scales. Note the weight differences given on each scale.
- Find the average of these differences and then divide this figure in half.
- This is the amount that has to be removed from one hold and relocated in the other hold, thus giving the overall effect.
- Where the mass is to be relocated from the rear hold to the fwd hold, from the vertical line raised from the current CG location into Cpt 4, count to the left the amount to be removed from this hold. Then raise a vertical Line into the fwd hold (Cpt 1).
- From this line, count to the left the new mass to be located in this hold (original + relocated). Then drop a vertical into Cpt 4.
- From this vertical, count to the right the new mass to be located in the rear hold (original – relocated). The last unit of this should coincide with the vertical raised from the new CG position.
- For cargo baggage relocations in the opposite direction, start at Cpt 1 and continue.

This system is used for passenger relocation.

Note: The relocation of masses does not alter the gross mass but alters all the CG locations.

SOLVING A SCALE SPACE PROBLEM

Where a load for a compartment is within the limiting values of the compartment but there is insufficient space on the row, the horizontal line would exit the scale. For example:

- DOI was 40.0
- 3000 kg for Cpt 1
- 4000 kg for Cpt 4

1. Enter the DOI, drop a vertical into Cpt 4, and add the scale component for the 4000 kg. Then raise a vertical into Cpt 1.

2. Add the scale component for Cpt 1 to this vertical and then drop the following vertical into Cpt Oa.

3. This can be done for any of the loads provided they are within the compartment's limits.

PRACTICE QUESTIONS BASED ON THE MRJT-1 DATA SHEET

Using the loading manifest and CG envelope on pages 30 and 31, answer the following five questions for an aircraft loaded as detailed below.

Aircraft MRJT	DOM	33 000 kg	CG at Stn 650
Crew	Standard		
Pay load	Pax 141		
	Baggage 282	Item at standard allowance loaded equally between holds	
Fuel load	Wing tanks full		
	Centre tank	600 kg	
	Start fuel	600 kg	
	Trip fuel	6000 kg	
Flap settings	Take-Off 15°		
	Landing 40°		

1. What is the BEM and CG for this aircraft?
   a. 32 580 kg at 21.56% MAC
   b. 32 580 kg at 21.33% MAC
   c. 32 550 kg at 21.56% MAC
   d. 32 550 kg at 21.33% MAC

2. What is the total payload for the aircraft?
   a. 15 510 kg
   b. 13 677 kg
   c. 11 844 kg
   d. 3666 kg

3. What is the aircraft's CG when it lands?
   a. 16.88% MAC
   b. 16.68% MAC
   c. 16.55% MAC
   d. 16.51% MAC

4. What is the stabilator trim required for take-off?
   a. 3.75
   b. 3.5
   c. 3.25
   d. 3.0

5. If there are no performance limitations for the flight, what is the underload?
   a. 29 800 kg
   b. 5489 kg
   c. 5149 kg
   d. 3306 kg

For the following two questions, use the Load and Trim sheet in CAP 696 for the aircraft as detailed below.

The aircraft has a DOM of 36 588 kg and a DOI of 50.0. It will make a flight to a destination XYZ where it is regulated to a landing mass of 50 900 kg. The trip fuel is 1950 kg but for contingency, reserve, and return fuel, the aircraft must land with 3000 kg of fuel remaining.

6. What is the allowed traffic load for this flight?
   a. 7012 kg
   b. 7412 kg
   c. 7400 kg
   d. 7000 kg

7. What is the aircraft's OM and CG?
   a. 15% MAC at 39 588 kg
   b. 15.25% MAC at 39 588 kg
   c. 15.5% MAC at 41 538 kg
   d. 15.75% MAC at 41 538 kg

8. An aircraft that has a DOM of 34 000 kg and DOI of 43.0 is to take a pay load of 60 adult male pax each with 13 kg of baggage and a cargo of 600 kg.

   The aircraft is performance limited to a MTOM of 57 000 kg and due to regulations at the destination, must land with 2000 kg of fuel remaining. The trip fuel is 7000 kg.
   The aircraft is loaded as follows:

   Baggage in Fwd hold
   Cargo in Aft hold
   12 Pax in Cpt Ob
   24 Pax in Cpt Oc
   24 Pax in Cpt Od

   From the following, choose the correct statements:
   i. The underload is 7580 kg
   ii. The traffic load is 14 000 kg
   iii. The ZFM is 40 420 kg
   iv. The baggage mass is 780 kg
   v. The Pax mass is 5060 kg
   vi. The ZFM CG is in limits
   vii. The fuel index is 0.1

   a. i, iii, iv, vii.
   b. i, ii, iv, v.
   c. ii, v, vi, vii.
   d. iii, iv, v, viii.

9. For an aircraft loaded as follows:

Aircraft	DOM 34 000 kg	DOI 43.0
Payload		
Pax at 84 kg	Cpt Ob 12	
	Cpt Oc 24	
	Cpt Od 24	
Baggage at 13 kg	Cpt 1 60 items	
Cargo	Cpt 4 600 kg	
Fuel	Bulk 10 000 kg	
	Start 1000 kg	
	Trip 7000 kg	

Calculate the ZFM CG as loaded. If the CG is out of limits, relocate the Pax in Cpt Od to Cpt Oe. Give the answer to the nearest % MAC.

a. 11%
b. 12%
c. 13%
d. 14%

10. For an aircraft that has a DOM of 33 470 kg and a DOI of 47.5, which is to make a flight where the only restrictions are those that are structural limitations.

The crewing is standard

The payload is the difference MZFM and DOM

Choose the correct statements for this flight.

i. Payload 17 830 kg
ii. Take-off fuel equivalent to a fuel index of – 5.7
iii. Take-off fuel equivalent to a fuel index of – 6.3
iv. Useful load for this flight is 29 330 kg
v. Useful load for this flight is 29 780 kg
vi. Mass of fuel in the centre tank at take-off is 2410 kg
vii. Mass of fuel in the centre tank at take-off is 2416 kg
viii. Mass of fuel available for start, run up and taxi is 260 kg
ix. Take–off fuel in US gallons is 349 001

a. ii, iii, vi, vii, ix.
b. i, ii, v, vii, viii.
c. i, iv, vi, vii, viii.
d. ii, iv, vi, vii, ix.

For questions 1-5

Max Permissible Aeroplane Mass Values:

TAXI MASS - ZERO FUEL MASS -

TAKE-OFF MASS - LANDING MASS -

ITEM	**MASS kg**	**BA IN**	**MOMENT kg in/1000**	**CG %MAC**
1. DOM				
2. PAX Zone A		284.00		-
3. PAX Zone B		386.00		-
4. PAX Zone C		505.00		-
5. PAX Zone D		641.00		-
6 PAX Zone E		777.00		-
7. PAX Zone F		896.00		-
8. PAX Zone G		998.00		-
9. CARGO HOLD 1		367.9		-
10. CARGO HOLD 4		884.5		-
11. ADDITIONAL ITEMS				-
ZERO FUEL MASS				
12. FUEL TANKS 1 & 2				
13. CENTRE TANK				
TAXI MASS				
LESS TAXI FUEL				
TAKE-OFF MASS				
LESS FLIGHT FUEL				
EST. LANDING MASS				

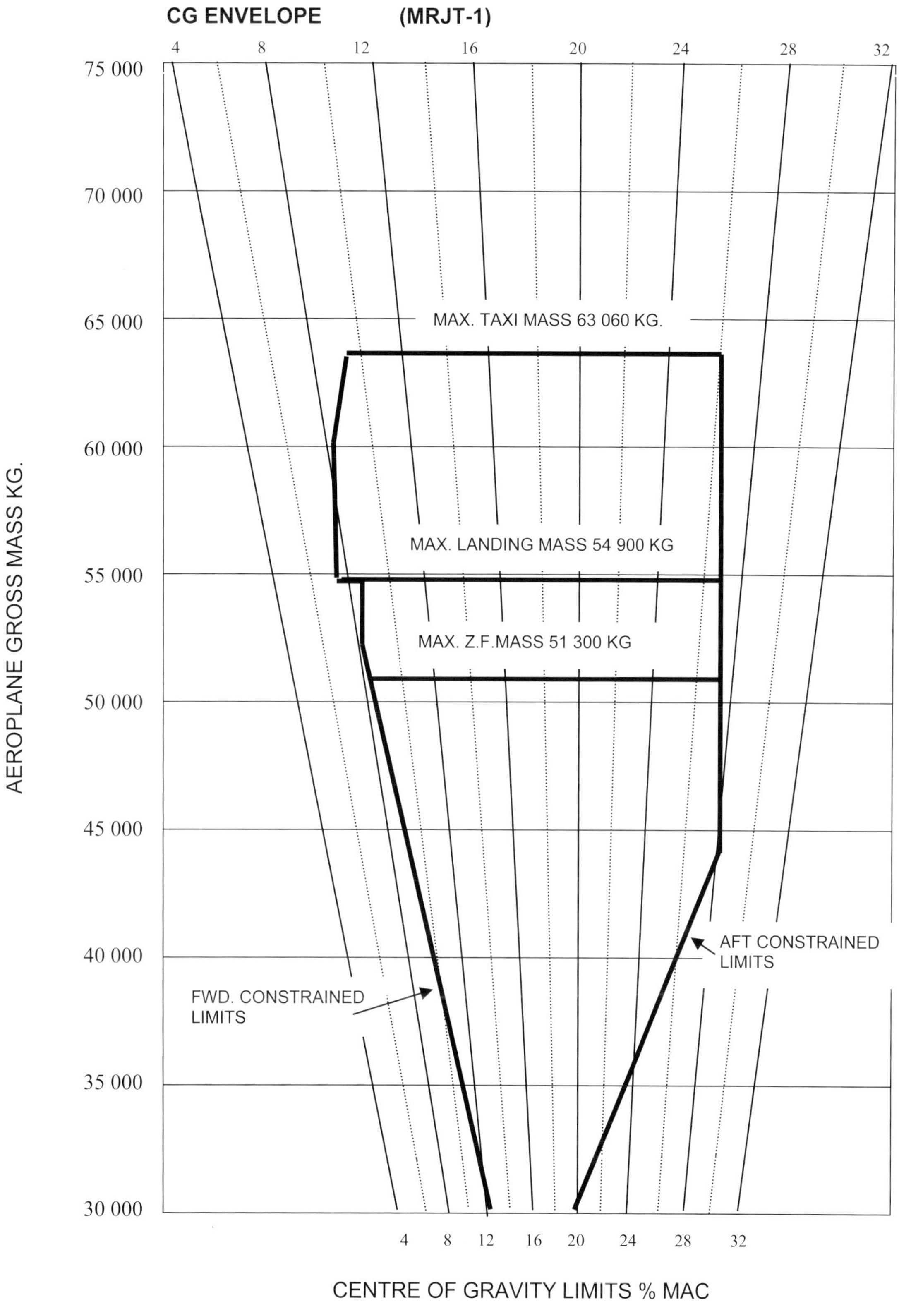
CG ENVELOPE (MRJT-1)
4 8 12 16 20 24 28 32
75 000
70 000
65 000
60 000
55 000
50 000
45 000
40 000
35 000
30 000
AEROPLANE GROSS MASS KG.
MAX. TAXI MASS 63 060 KG.
MAX. LANDING MASS 54 900 KG
MAX. Z.F.MASS 51 300 KG
FWD. CONSTRAINED LIMITS
AFT CONSTRAINED LIMITS
4 8 12 16 20 24 28 32
CENTRE OF GRAVITY LIMITS % MAC

ANSWERS TO PRACTICE QUESTIONS

1.	B
2.	C
3.	B
4.	C
5.	D
6.	B
7.	D
8.	A
9.	B
10.	B

OTHER METHODS OF DERIVING THE CG'S LOCATION

There are several further methods of working out an aircraft's CG. These methods do not have to be learnt, but be aware of their existence.

COMPUTER GENERATED LOAD AND TRIM SHEETS

All data is entered and the computer calculates and prints the load and trim sheet.

SLIDE RULES

There are dedicated slide rules for finding the CG and GM for various conditions and calculating the effect of load transfer, fuel consumption, etc., and are either rotary or linear.